Recent developments in laboratory identification techniques

Recent developments in laboratory identification techniques

Proceedings of a symposium
Uppsala
September 19–20, 1979

Editors

R. Facklam
G. Laurell
I. Lind

 1980

Excerpta Medica, Amsterdam-Oxford-Princeton

International Congress Series 519
ISBN Excerpta Medica 90 219 9438 0
ISBN Elsevier North-Holland 0 444 90152 3

Library of Congress Cataloging in Publication Data

Main entry under title:

Recent developments in laboratory identification techni-
    ques.

    (International congress series ; 519)
    Includes bibliographies.
    1. Bacteria, Pathogenic--Identification--Congresses.
2. Diagnosis, Laboratory--Congresses. 3. Neisseria
gonorrhoeae--Identification--Congresses. 4. Streptococ-
cus--Identification--Congresses. I. Facklam, R.
II. Laurell, G. III. Lind, Inga. IV. Series.
[DNLM: 1. Neisseria gonorrhoeae--Isolation and puri-
fication--Congresses. 2. Streptococcaceae--Isolation
and purification--Congresses. 3. Bacteriological
technics--Congresses. W3 EX89 no. 519 1979 / QW131 R295
1979]
QR67.R42   616'.014'028   80-15619
ISBN 90-219-9438-0 (Excerpta Medica)
ISBN 0-444-90152-3 (Elsevier North-Holland)

Publisher:
Excerpta Medica
305 Keizersgracht
P.O. Box 1126
1000 BC   Amsterdam

Sole distributors for the USA and Canada:
Elsevier North-Holland Inc.
52 Vanderbilt Avenue
New York, N.Y. 10017

Printed in The Netherlands by Dijkstra Niemeyer b.v.

Contents

Welcome

G. Laurell

It is a privilege for me to welcome so many prominent scientists to this conference. Before we start I should like to say a few words about the two bacteria, gonococci and streptococci, that we are going to discuss during these two days. It is quite obvious that there has been tremendous interest in all fields of research work regarding gonococci. The National Institute of Allergy and Infectious Diseases has initiated a number of investigations and the amount of money spent on gonococci has increased from 75,000 dollars in 1970 to 4 million dollars in 1975. There are many explanations for this but two are probably more important than others.

There has been an increase of gonococcal infections in many countries. Over one million cases were reported to the U.S. Public Health Service in 1976. In some countries, such as, for example, Sweden, there has been a decrease. Such differences are difficult to explain but might be due to under-reporting. It is, however, obvious that among the sexually transmitted diseases gonococcal infection is still the most important. Interest increased with the recognition of penicillinase-producing strains of gonococci in the Philippines late in 1975 and in the United Kingdom and the United States early in 1976, this at a time when the number of infections was increasing in several countries. After that first recognition, penicillinase-producing gonococci have been isolated in a large number of countries, including Sweden. It is no wonder that public health people became worried. Since we had effective antibiotics we had up to that time thought that we could control or even eradicate the disease without worrying too much about epidemiology, virulence, pathogenesis and so on. It appeared to be more a social than a microbiological problem. As so many times before, nature set us right and it is quite clear that we must go much deeper into the characteristics of gonococci than we thought. This meeting will deal only with small but very important parts of this complex field, namely, the transport of specimens to the laboratory, suitable media for cultivation and fast and reliable methods for identifying gonococci. If we do not master these, it will be difficult to determine the effect of measures such as prophylaxis, antibiotic treatment, future vaccines and so on.

On the second day of this meeting diagnostic problems relating to isolation, identification and typing of β-haemolytic streptococci will be discussed. Streptococci are a very large group of bacteria which cause infections not only in man but also in cattle, horses, pigs, sheep and other animals. It will not be possible to discuss problems connected with disease among all these animals. As far as I can see from the programme the speakers will talk mainly about β-haemolytic streptococci causing disease in man. These belong to a fairly limited number of groups within the Lancefield system. Contrary to the situation with respect to gonococci, well established typing systems, serological and others, exist and the panorama of infection is well known. Thus it is well known, for example, that there has been a decline in rheumatic fever in Western Europe and the United States that started long before the antibiotic era. In Scandinavia, rheumatic fever is now rare, as is puerperal sepsis. In hospital infections such as postoperative wound infections, streptococci are no longer a great problem, at least not in Sweden. Exceptions are infected burns and intermittent smaller outbreaks in dermatological wards. On the other hand, streptococcal infections continue to be one of the most frequent bacterial diseases in the community. In the literature reports of serious outbreaks caused by single virulent types usually rich in M protein may also be seen now and then. The consequences of streptococcal infections, such as rheumatic fever and nephritis, are well known but little is known about why some streptococci in some people initiate either of these diseases.

Some haemolytic streptococci with rather weak haemolysis belong to group B and infections caused by these have aroused much interest in the last five or six years. This is explained by the fact that neonatal infections secondary to group B streptococci represent one of the most fulminant infections known to man. In many series, group B streptococci represent one of the two most common bacterial pathogens in neonatal infections. It has been estimated that two to three per 1000 live neonates develop an infection with these organisms: one per 1000 dies. At a conference in the United States in 1977 it was estimated that between 12,500 and 15,000 babies would develop the disease during 1978. Approximately 50% of these would die and up to 50% of the survivors with meningitis would develop neurological sequelae.

We now have effective treatment with penicillin and can therefore save the patients if the infections are discovered early enough, but for how long? I would like here to cite Sir Robert Williams who, at the Oxford symposium on streptococcal infections in 1978, said: "Will streptococci at long last pick up some genes for penicillin resistance? The gonococci have shown that it is never too late to learn".

It is no wonder that we are interested in how to transport, cultivate, identify and type the streptococci. Many researchers have worked on this problem during the last 70 years and we have a large number of biochemical and immunological methods at our disposal. Many of them function reasonably well in the hands of experienced people but are often too slow when it comes to diagnosing life-threatening diseases like meningitis. I hope that the speakers will tell us what there is to be seen for gonococci and streptococci in the crystal ball: immunofluorescence, co-agglutination or other new methods. We need these. After these few introductory words it is a pleasure to open this meeting and I wish it all success.

I

Neisseria gonorrhoeae

The laboratory diagnosis of gonorrhoea

I. Lind
WHO Collaborating Centre for Reference and Research in Gonococci, Neisseria Department, Statens Seruminstitut, Copenhagen, Denmark

Gonorrhoea is one of the most common infectious diseases all over the world. The prevalence is high even in developed countries with the most sophisticated public health services, including treatment free of charge. As an example, the prevalence of gonorrhoea in Denmark during the last 20 years is illustrated in the Figure.

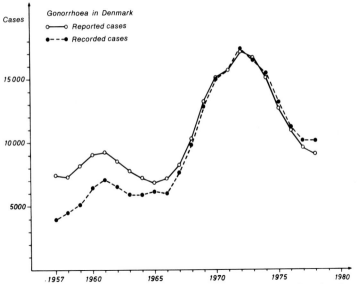

Figure. Gonorrhoea in Denmark. o — o : *annual number of cases reported to the National Health Service;* ● - - - ● : *annual number of cases recorded at the Neisseria Department on the basis of positive culture for Neisseria gonorrhoeae.*

From 1972 to 1977, the annual number of cases decreased from about 17,000 to about 10,000. This decrease stopped in 1977/78 and the preliminary figures for 1979 show a slight increase. For many years the annual number of cases reported to the health authorities has been about the same as that recorded at the Neisseria Department on the basis of positive culture for gonococci. In 1978, the number of cases recorded exceeded the number reported by about 10 %. The total population of Denmark is about five million. This means that the morbidity is about 200 cases per 100,000 inhabitants. For the highly exposed age-groups between 15 and 25 years of age, the morbidity is up to 2,000 per 100,000 or 1–2 %.

3

Table I. Sensitivity and specificity of a diagnostic test.

$$\text{Sensitivity} = \frac{\text{Diseased persons with positive test}}{\text{All diseased persons tested}} \times 100$$

$$\text{Specificity} = \frac{\text{Non-diseased persons negative to the test}}{\text{All non-diseased persons tested}} \times 100$$

The figures do not cover the northern part of Denmark, the Arctic island of Greenland. The problem is much more serious in that part of the country. As in Denmark, about 10,000 cases are reported each year but the total population is only 50,000. Thus the prevalence of gonorrhoea in Greenland is about 100 times higher than that in Denmark. Since the reliability of a diagnostic test is dependent on the prevalence of the corresponding disease [1], a primary consideration in our selection of methods for the laboratory diagnosis of gonorrhoea is whether they are to be applied to the population in Greenland or in Denmark. The methods have also to meet a number of minimum requirements. The reliability of a test in distinguishing diseased from non-diseased persons is often defined by its sensitivity and specificity [1]. Sensitivity is the ability of a test to give a positive finding when the person tested truly has the disease under study. Specificity is the ability of the test to give a negative finding when the person tested is free from the disease under study (Table I).

When a diagnostic test is to be used in a large, unselected population, it is important to know the likelihood that a subject yielding a positive test actually has the disease – i.e., the predictive value of a positive test result (Table II). The predictive value of a negative test result is the likelihood that a subject with a negative test does not have the disease.

Predictive values cannot be estimated directly from sensitivity and specificity since they are dependent on the prevalence of the disease in the population studied. This is illustrated by calculation of the predictive value of positive test results for two groups of individuals, one in which the prevalence of the disease is 30 % and one in which the prevalence is 1 % (Table III). The test has a sensitivity of 80 % and a specificity of 90 %. The predictive values of a positive test are 77 and 8, respectively. As indicated by the figures in brackets, the predictive value of positive test results is improved when the specificity is improved to 98 %.

The minimum requirements for an adequate screening test for gonorrhoea have been analyzed by Galen and Gambino [2]. They conclude that the sensitivity ought to be greater than 80 % and the specificity greater than 99 % while the test must be capable of differentiating current from previous infection.

Table II. Predictive values of a diagnostic test.

$$\text{Predictive value of positive test} = \frac{\text{Number of diseased persons with positive test}}{\text{Total number of persons with positive test}} \times 100$$

$$\text{Predictive value of negative test} = \frac{\text{Number of non-diseased persons with negative test}}{\text{Total number of persons with negative test}} \times 100$$

Table III. Predictive values of a diagnostic test in relation to incidence of disease. Test with 80 % sensitivity and 90 (98) % specificity.

	Predictive values	
	1,000 individuals, 30 % incidence of disease	1,000 individuals, 1 % incidence of disease
Positive test results	77 (94) %	8 (29) %
Negative test results	91 (92) %	100 (100) %

The methods available for the laboratory diagnosis of gonorrhoea can be divided into three groups, namely, methods for the demonstration of gonococci or gonococcal components in specimens (Table IV), cultural procedures and serological tests.

The most simple and rapid methods are staining of smears with methylene blue or according to Gram's method. The advantages and drawbacks of these methods are well known [3, 4]. The recently developed techniques for antigen detection have so far been evaluated only by examination of specimens from highly selected groups of patients [5–11]. All the immunological methods are hampered by the facts that gonococci show pronounced variation in their antigenic structure, that the amount of antigen present varies, and may be low, especially in female patients, and that locally-produced antibodies may interfere with the test. Methods based on immunological detection of gonococcal antigens directly in specimens cannot, therefore, be expected to be sensitive enough for diagnostic purposes. Perhaps some of the later speakers will approach the matter from another angle.

Available serological tests do not distinguish between present and past infection, resulting in low values for the predictive value of positive test results. A serological test for the diagnosis of existing gonococcal infection seems, therefore, to be unattainable, although tests with improved sensitivity and specificity have recently been developed [12–14].

Cultural procedures are, consequently, still the first choice for the laboratory diagnosis of gonorrhoea. The predictive value of a positive test is 100 %, specificity is 100 % and the sensitivity is reported to be between 70 and 95 % [4, 15].

Factors to be considered in relation to cultural procedures are listed in Table V. Problems involved in the establishment of cultural procedures will no doubt be discussed in other papers. I wish to point out only that the sensitivity of cultural procedures for the diagnosis of gonorrhoea is far more dependent on, for example,

Table IV. Direct identification of gonococci or gonococcal components in specimens.

Stained smears:
  Methylene blue
  Gram's method
  Acridine orange
  Fluorescent antibody technique
Limulus assay
Genetic transformation
Counterimmunoelectrophoresis
Radioimmunoassay

Table V. Cultural procedures for the laboratory diagnosis of gonorrhoea.

1) Sampling sites
2) Taking of specimens
3) Transport media
4) Growth media
5) Incubation conditions
6) Identification of isolates

Table VI. Percentage of patients with gonorrhoea from whom pharyngeal specimens were received for bacteriological examination.

Year	Females (%)	Males (%)
1972	10	10
1973	17.9	22.8
1974	21.5	29.5
1975	23.2	32.4
1976	23.2	34.2
1977	22.7	35.9
1978	24.3	40.5

the media used, than on the identification methods which are available, while the sampling sites are also important from an epidemiological point of view. It is often discussed, in Denmark at least, whether it is reasonable to take pharyngeal and rectal specimens in addition to urethral and/or cervical specimens.

In Denmark, pharyngeal gonorrhoea was rediscovered in 1971 or 1972 [16]. Since then an increasing proportion of patients has had pharyngeal specimens taken [17]. In 1978 pharyngeal specimens were taken from about 25 % of all female and 40 % of all male patients with gonorrhoea (Table VI). A total of about 250 cases of pharyngeal gonorrhoea is diagnosed each year (Table VII). This figure corresponds to a frequency of 10–12 % among female patients and 7–8 % among male patients examined for pharyngeal gonorrhoea. The figures for 1979 appear to be even higher.

Table VII. Pharyngeal gonorrhoea in Denmark. Figures in brackets indicate the number of patients in whom the pharynx was the only positive site.

Year	Females	Males	Total
1971	28 (2)	44 (12)	72
1972	120 (?)	110 (?)	230
1973	125 (36)	118 (44)	243
1974	132 (40)	149 (63)	281
1975	109 (30)	112 (41)	221
1976	127 (24)	102 (49)	229
1977	111 (25)	154 (46)	265
1978	147 (39)	126 (51)	273

Table VIII. Rectal gonorrhoea in Denmark (rectum the only positive site).

Year	Females	Males	Total
1972	124	132	256
1973	140	95	235
1974	112	108	220
1975	73	132	205
1976	77	102	179
1977	58	146	204
1978	53	111	164

Table VIII shows the number of patients with isolated rectal gonorrhoea during the same period. In some clinics the taking of pharyngeal specimens has replaced taking of rectal specimens. It is suggested, therefore, that the numbers recorded in Table VIII are too low.

The effect of the medium is probably the most critical factor in the cultural diagnosis of gonorrhoea. During the last eight years each specimen received at the Neisseria Department has been inoculated on two chocolate agar media, one selective medium containing nystatin, polymyxin and vancomycin and a so-called nonselective medium to which only trimethoprim is added [3]. The recovery of gonococci through culture on one or both media for 2,701 isolates in 1978 is shown in Table IX [17], from which it may be seen that the recovery from rectal and more especially pharyngeal specimens is markedly improved by the use of selective medium. About 120 isolates (4–5%) were recovered only on the nonselective medium. It has previously been shown that 3–4% are strains which are inhibited by vancomycin at the concentration used in our medium, 2 µg/ml. From a cost-benefit point of view, the advantage of using two media, or rather of including a nonselective medium, was questioned by a regional laboratory, which omitted the nonselective medium for a period of two to three years. During that period the prevalence of vancomycin-sensitive strains in this region increased from about 3% to 15%. This means that the composition of the medium alone can cause a decrease in the sensitivity of cultural procedures to 85% or less.

However, the number of specimens received in our laboratory has shown a

Table IX. Recovery of Neisseria gonorrhoeae by culture on selective and/or nonselective media from urogenital, rectal* and pharyngeal specimens.

	Number of isolates recovered from			Total
	Urethra/cervix	Rectum	Pharynx	
Both types of media	2,370	55	69	2,504
Nonselective medium only	117	1	1	119
Selective medium only	77	20	79	176
Total	2,564	76	149	2,701

* The figures comprise only patients in whom the rectum was the only positive site

constant increase. This promoted an interest in improving the selective medium. For the last two months a medium containing lincomycin (2 μg/ml) instead of vancomycin (2 μg/ml) has been evaluated. Out of 2,000 isolates five were recovered only on the nonselective medium. By subculture on the lincomycin-containing medium all five strains showed a normal degree of growth. This medium is, therefore, regarded as very suitable for the isolation of gonococcal strains now present in Denmark, at least.

The final step in the cultural procedures, the identification of isolates, is our main subject today.

Table X. Identification of N. gonorrhoeae isolates.

Presumptive diagnosis:
Colony morphology
Oxidase reaction

Confirmatory diagnosis:
a) Bacteriological identification:
Gram-stained smears
Carbohydrate utilization test
Genetic transformation
Enzymatic profiles

b) Immunological identification:
Direct immunofluorescence test
Co-agglutination test
Counterimmunoelectrophoresis
Slide agglutination test

The identification procedure (Table X) starts with inspection of the inoculated medium and the establishment of a presumptive diagnosis based on colony morphology and a positive oxidase reaction. The methods available for confirmatory diagnosis can be divided into two groups, bacteriological [6, 18–22] and immunological [8, 23–28]. Those in common use are italicized. In our laboratory the direct immunofluorescence test is used routinely. I should like to demonstrate the reliability of the presumptive diagnosis by means of some previously published data [26]. During a period of two months, specimens from 976 male patients showed growth of gonococcal-like oxidase-positive colonies. Of these 26, or 2.7 %, were negative in the direct immunofluorescence test. For female patients 82, or 6 %, of 1,372 gonococcal-like isolates were negative. The predictive value of the presumptive diagnosis (colony morphology and oxidase reaction) was thus about 95 % when applied to specimens received from various low- and high-risk groups of patients. In laboratories which only receive specimens from high-risk groups, e.g., in Greenland, there is no need for further identification procedures.

This means that the debate as to further identification procedures turns on problems which only arise in developed countries or those with a fairly low prevalence of gonorrhoea.

REFERENCES

1. Vecchio, T.J. (1966): Predictive value of a single diagnostic test in unselected populations. *N. Engl. J. Med., 274*, 1171.
2. Galen, R.S. and Gambino, S.R. (1975): Screening for gonorrhoea. How the model defines essential criteria for an efficient screening test. In: *Beyond Normality: The Predictive Value and Efficiency of Medical Diagnosis.* John Wiley & Sons, New York.
3. Reyn, A. (1965): Laboratory diagnosis of gonococcal infections. *Bull. W.H.O., 32*, 449.
4. WHO Scientific Group (1978): *Neisseria gonorrhoeae* and gonococcal infections. *Technical Report Series 616.* World Health Organization, Geneva.
5. Forsum, U. and Hallen, A. (1978): Acridine orange staining of urethral and cervical smears for the diagnosis of gonorrhoea. *Acta Derm.-Venereol., 59*, 281.
6. Janik, A., Juni, E. and Heym, G.A. (1976): Genetic transformation as a tool for detection of *Neisseria gonorrhoeae. J. Clin. Microbiol., 4*, 71.
7. Kwapinski, G., Kwapinski, E. and Spurrier, M. (1976): Identification of gonococcal cultures by means of their cytoplasmic antigens and an anticytoplasm antiserum. *Health Lab. Sci., 13*, 262.
8. Lind, I. (1967): Identification of *Neisseria gonorrhoeae* by means of fluorescent antibody technique. *Acta Pathol. Microbiol. Scand., 70*, 613.
9. Spagna, V.A., Prior, R.B. and Perkins, R.L. (1979): Rapid presumptive diagnosis of gonococcal urethritis in men by the limulus lysate test. *Br. J. Vener. Dis., 55*, 179.
10. Thornley, M.J., Wilson, D.V., Demarco de Hormaeche, R., Oates, J.K. and Coombs, R.R.A. (1978): Detection of gonococcal antigens in urine by radioimmunoassay. *J. Med. Microbiol., 12*, 161.
11. Thornley, M.J., Andrews, M.G., Briggs, J.O. and Leigh, B.K. (1978): Inhibitors in urine of radioimmunoassay for the detection of gonococcal antigens. *J. Med. Microbiol., 12*, 177.
12. Buchanan, T.M., Swanson, J., Holmes, K.K., Kraus, S.J. and Gotschlich, E.C. (1973): Quantitative determination of antibody to gonococcal pili. Changes in antibody levels with gonococcal infection. *J. Clin. Invest., 52*, 2896.
13. Reimann, K. and Lind, I. (1977): An indirect haemagglutination test for demonstration of gonococcal antibodies using gonococcal pili as antigen. I. Methodology and preliminary results. *Acta Pathol. Microbiol. Scand. Sect. C, 85*, 115.
14. Reimann, K., Lind, I. and Andersen, K.E. (1980): An indirect haemagglutination test for demonstration of gonococcal antibodies using gonococcal pili as antigen. II: Serological investigation of patients attending a dermato-venereological out-patient clinic in Copenhagen. *Acta Pathol. Microbiol. Scand. Sect. C*, in press.
15. Danielsson, D., Sandström, E., Kjellander, J., Moi, H. and Wallmark, G. (1978): Culture diagnosis of gonorrhoea – a comparison between two standard laboratory methods and a commercial gonococcal culture kit (Kvadricult®). *Acta Derm. Venereol., 58*, 69.
16. Bro-Jørgensen, A. and Jensen, T. (1973): Gonococcal pharyngeal infections. Report of 110 cases. *Br. J. Vener. Dis., 49*, 491.
17. Lind, I., Bollerup, A.C. and Reimann, K. (1978): Activities of the WHO Collaborating Centre for Reference and Research in Gonococci, Copenhagen, for the year 1978. *WHO/VDT/RES/GON/79, 124.* World Health Organization, Geneva.
18. Bawdon, R.E., Juni, E. and Britt, E.M. (1977): Identification of *Neisseria gonorrhoeae* by genetic transformation: a clinical laboratory evaluation. *J. Clin. Microbiol., 5*, 108.
19. D'Amato, R.F., Eriques, L.A., Tomfohrde, K.M. and Singerman, E. (1978): Rapid identification of *Neisseria gonorrhoeae* and *Neisseria meningitidis* by using enzymatic profiles. *J. Clin. Microbiol., 7*, 77.
20. Van Dyck, E., Piot, P. and Pattyn, S.R. (1976): Evaluation of three techniques for carbohydrate fermentation of Neisseriae. *Zentralbl. Bakt. Parasitenkd. Infektionskr. Hyg. Abt. 1 Orig. A, 236*, 231.

21. Morse, S.A. and Bartenstein, L. (1975): Adaptation of the Minitek System for the rapid identification of *Neisseria gonorrhoeae*. *J. Clin. Microbiol.*, *3*, 8.
22. Shtibel, R. and Toma, S. (1977): *Neisseria gonorrhoeae*: evaluation of some methods used for carbohydrate utilization. *Can. J. Microbiol.*, *24*, 177.
23. Arko, R.J., Wong, K.H. and Peacock, W.L. (1979): Nuclease enhancement of specific cell agglutination in a serodiagnostic test for *Neisseria gonorrhoeae*. *J. Clin. Microbiol.*, *9*, 517.
24. Barnham, M. and Glynn, A.A. (1978): Identification of clinical isolates of *Neisseria gonorrhoeae* by a co-agglutination test. *J. Clin. Pathol.*, *31*, 189.
25. Danielsson, D. and Kronvall, G. (1974): Slide agglutination method for the serological identification of *Neisseria gonorrhoeae* with anti-gonococcal antibodies adsorbed to protein A-containing staphylococci. *Appl. Microbiol.*, *27*, 368.
26. Lind, I. (1975): Methodologic aspects of routine procedures for identification of *Neisseria gonorrhoeae* by immunofluorescence. *Ann. N.Y. Acad. Sci.*, *254*, 400.
27. Malysheff, C., Wallace, R., Ashton, F.E., Diena, B.B. and Perry, M.B. (1978): Identification of *Neisseria gonorrhoeae* from primary cultures by a slide agglutination test. *J. Clin. Microbiol.*, *8*, 260.
28. Menck, H. (1976): Identification of *Neisseria gonorrhoeae* in cultures from tonsillo-pharyngeal specimens by means of a slide co-agglutination test (Phadebact® Gonococcus Test). *Acta Pathol. Microbiol. Scand. Sect. B*, *84*, 139.

Discussion

S. A. Waitkins (Liverpool, England): When lincomycin is used in the selective medium, does Dr Lind find that the contamination rate increases and that the method, therefore, perhaps becomes slightly more difficult for the inexperienced technician?

I. Lind: No, the contamination rate decreases. For the examination of pharyngeal specimens the new medium is definitely better than the old one containing vancomycin. Initially, we tried to omit vancomycin as well as lincomycin from the selective medium. This resulted in the loss of the majority of gonococcal isolates from pharyngeal and rectal specimens.

G. Laurell (Uppsala, Sweden): I have seen some published reports that gonococci causing severe diseases, such as sepsis, are more sensitive to penicillin than those causing less severe diseases. Is this true?

I. Lind: We have no data from Denmark. I do not know.

G. Laurell: Is it also true that plasmids for penicillinase production could be just as easily transferred to these more virulent strains as to the less virulent strains that are isolated from the urethra?

I. Lind: There is no direct evidence for that. However, the strains which lose their plasmids coding for β-lactamase production, for example, by lyophilization – as can happen – show a susceptibility to penicillin anywhere within the entire range known for gonococci. They may be sensitive, less sensitive or resistant. It seems, therefore, that the β-lactamase plasmid is acquired by different kinds of gonococcal strains.

J. A. Morello (Chicago, U.S.A.): Does Dr Lind know whether any of the patients from whom penicillinase-producing resistant strains have been isolated have been capable of being successfully treated with penicillin? Are these patients resistant to treatment?

I. Lind: We have had one such case cured through standard penicillin treatment.

J. A. Morello: So these cases are picked up because of routine screening for penicillinase production. By which test is the penicillinase measured?

I. Lind: We use the chromogenic cephalosporin test. Initially, we used two tests, a rapid iodometric test and the chromogenic cephalosporin test. Since the results obtained were identical, we have chosen the simplest for routine use.

R. R. Facklam (Atlanta, U.S.A.): I am not a gonococcus person but I am interested in Dr Lind's immunofluorescence technique. How many immunofluorescence tests are done in a day by one technician?

I. Lind: There are up to 100 isolates to be tested each day and the immunofluorescence test is performed by two technicians. We prefer one technician to do no more than about 50 in a day because she gets tired making the readings.

R. R. Facklam: Have any controls ever been done on the stability or accuracy of identification by the technicians towards the end of the day after looking at 200 immunofluorescence slides?

I. Lind: No, they never do more than 50 in a day.

R. R. Facklam: I think it would be a very tedious type of procedure to give somebody to do. At least, that has been our experience in the States among the people who look at the immunofluorescence slides. Hundreds of thousands of slides are done for Group A streptococci. One of the major problems with immunofluorescence is technical error towards the end of the day. I suspect that if it were possible to devise a scheme or identification procedure that did not require someone to look through a microscope, it would be better.

I. Lind: We have not met such problems with our technique. Our microscopes are of very high quality – that is important.

G. Laurell: My own experience during the war when there was quite a lot of diphtheria in Sweden was of direct staining. Some of us had to look at 500 or 600 slides a day. This was very tiring and, it is quite true, not very reliable towards the end of the day.

J. Rotta (Prague, Czechoslovakia): I do not work in this field, but I wonder about the diagnostic or identification value of the limulus test in view of the recent finding that this test is also positive in patients with gonorrhoea?

I. Lind: I have seen one paper on that assay. The authors reported results only on specimens from male patients with symptoms of urethritis. They obtained the same results by the limulus test as those obtained by Gram-stained smears, i.e., the diagnosis was established in 95 % of cases.

P. Olcén (Örebro, Sweden): Is the immunofluorescence technique used solely on throat specimens to determine a gonococcal diagnosis?

I. Lind: No, the immunofluorescence test is used only for screening of the pharyngeal isolates in order to eliminate further work with oxidase-positive rods and the apathogenic Neisseria. The diagnosis is confirmed by carbohydrate utilization tests and Gram-stained smears.

J. Maeland (Trondheim, Norway): Could Dr Lind briefly state what is the composition of the medium at present used for primary isolation?

I. Lind: I cannot give the detailed composition of the chocolate agar medium, which is prepared at our Institute. An important ingredient is pork liver autolysate. The antibiotics added are polymyxin (25 units/ml), amphotericin B (2 μg/ml), trimethoprim (3 μg/ml) and, for the time being, lincomycin (2 μg/ml).

J. Vandepitte (Leuven, Belgium): What is Dr Lind's experience in her Institute with regard to receiving strains through the post from a long way away? For example, do strains from Morocco arrive in reasonable condition?

I. Lind: There are two possible ways of sending such strains. Firstly, the sample may be lyophilized and sent without any problems. Secondly, samples may be sent in liquid nitrogen.

J. Vandepitte: That method is probably not practicable for most parts of Africa.

I. Lind: We have received specimens from both Nigeria and the Far East in liquid nitrogen.

J. Vandepitte: That is all right when it is allowed but, in my experience, the Customs do not always permit specimens to be sent in this way.

I. Lind: The WHO has a special arrangement with most airline companies. So far, we have not met any difficulties with Customs.

J. S. Lewis (Atlanta, U.S.A.): Are there any plans for the control of gonorrhoea in Greenland?

I. Lind: There have been many, but they do not work! Epidemiological treatment has also been attempted in a restricted area, but three months later the prevalence increased again. It is primarily a social problem, closely connected with alcohol problems.

W. R. Maxted (London, England): Did I hear correctly that the gonococci lost a plasmid after lyophilization?

I. Lind: Yes.

W. R. Maxted: Is this a common occurrence among bacteria? We have seen some curious things happening among streptococci after lyophilization. Of course, it is very serious, if true.

I. Lind: Dr Rosendal at our Institute has told me that it is a rather common event in strains of *Staphylococcus aureus*. The plasmid-determined multiresistance is often lost during storage or lyophilization.

W. R. Maxted: The changes we have seen are antigenic changes which some people say are plasmid-associated. This is, of course, very serious for a reference laboratory.

13

T. Horodniceanu (Paris, France): In our experience with streptococci strains there has been no loss of plasmid by lyophilization.

I. Lind: How many strains have been lyophilized? We have tested about 200 β-lactamase-producing gonococcal strains before and after lyophilization and three lost the plasmid.

T. Horodniceanu: We have lyophilized about 50 strains over two or three years, with nos plasmid loss.

I. D. Amirak (London, England): What is the routine sensitivity test for the gonococci isolated?

I. Lind: It is a plate dilution method using fourfold dilution steps.

M. Veron (Paris, France): Is it difficult to distinguish between *Neisseria gonorrhoeae* and *Neisseria meningitidis* with throat specimens?

I. Lind: No, not so far. For this purpose I think that the quality of the test used for the determination of carbohydrate utilization is important.

I. D. Amirak: Which sugars are included in the carbohydrate utilization test?

I. Lind: In our routine procedure for pharyngeal specimens we use glucose, maltose and sucrose.

Clinical considerations and the need for identification of *Neisseria gonorrhoeae*

A. Lassus
Department of Dermatology, University Central Hospital, Helsinki, Finland

Gonorrhoea is an epidemiologically uncontrollable disease for two major reasons. First, most infected females and an increasing proportion of infected males are asymptomatic carriers. Secondly, there is the difficulty of making a correct, rapid clinical diagnosis in the female.

Gonorrhoea and other sexually transmitted diseases are essentially behavioural diseases and if sexual behaviour can be controlled so will venereal diseases. The national and international preoccupation with sex and sexual permissiveness hardly presages a change in sexual mores. Venereal disease as well as sex education courses in schools will not cause any significant change in behaviour but they will encourage patients to go to a physician or clinic earlier. This means an increase in symptomatic individuals suspecting a sexually transmitted disease, usually gonorrhoea, visiting physicians and clinics. This naturally helps, partly, to control the spread of the disease and to prevent complications but depends, however, on correct diagnosis.

An attack of gonorrhoea confers no immunity to subsequent reinfections, often caused by the same untreated partner. The phenomenon of re-exposure followed by reinfection, successful treatment and repetition of the cycle has been called 'ping-pong' gonorrhoea and ceases only when the infected consort is found, when diagnosis is correct and treatment is successful.

CLINICAL PICTURE

Symptoms of a gonococcal infection can occur as soon as one day or as late as two weeks or more following sexual contact. The average incubation period for males is three to five days. In females it is difficult, if not impossible, to know when symptoms first begin.

The diagnosis of acute gonorrhoea in the male presents little or no problem to the physician. There is a history of sexual exposure within the previous two weeks. The earliest symptoms are uncomfortable sensations in the urethra followed by frequency of urination. By this time the gonococcus has penetrated the columnar epithelium of the anterior urethra and gone into the submucosa. An inflammatory response is triggered, manifested by a mucoid urethral discharge with dysuria. This is followed, usually in a matter of hours, by a purulent urethral discharge, dirty yellow in colour. Constitutional symptoms, if present, are mild. The infection is localized to the

15

anterior urethra for the first two weeks or so and then spreads backwards to the posterior urethra, involving the prostate and often the seminal vesicles. From here the infection follows the vas to the epididymis, resulting in a painful, usually unilateral epididymitis.

In the female, acute gonorrhoea involves the urethra, Skene's glands, Bartholin's glands and cervix. The vagina is never affected after the age of puberty. Symptoms are therefore referable to the structures involved. However, the majority of females have no subjective symptoms of infection. Pelvic inflammatory disease may follow immediately after an acute infection or may be delayed for several months, though it is not seen, as a rule, before puberty or in the postmenstrual state. When it occurs, both Fallopian tubes are frequently affected. The resulting scar tissue may block the lumen or trap the purulent discharge to form a pyosalpinx which may in turn provoke a peritonitis.

Other complications can also occur. Proctitis, when it arises in males, is almost always a result of homosexual contact but in women it is frequently caused by direct spread from the vaginal discharge as well as genitorectal exposures. Pharyngeal gonorrhoea is often asymptomatic but may occasionally present itself as a sore throat. Positive gonococcal pharyngeal cultures correlate positively with admitted practice of fellatio. Bro-Jørgensen and Jensen [1], in a series of 1,152 patients with gonorrhoea, found 10% of the women, 7% of the heterosexual males and 25% of the homosexual males to have pharyngeal infection. Ödegaard and Gunderson [2], in a series of 1,440 consecutive patients with gonorrhoea, found the infection only in the pharynx of about 1%. Willmott [3] has suggested that gonococci may be passed directly from throat to throat. Regardless of the clinical implications of a positive gonococcal pharyngeal culture, it is certain that a proportion of patients with urogenital gonorrhoea also carry gonococci in their pharynx. These gonococci are frequently not eradicated by routine treatment of genital gonorrhoea.

Gonococcal ophthalmia is predominantly a preventable neonatal problem. Gonococcal conjunctivitis in adults is rare and usually acquired by autoinoculation. Babies' eyes are infected by the mother whose secretions contain gonococci. These organisms usually come from infected cervical glands, some of which are opened up by cervical dilatation and the pressure of the child as it makes its way through the birth canal. Gonococcal ophthalmia neonatorum has been reported, though rarely, to have developed in spite of Credé's prophylaxis.

Gonorrhoeal arthritis, which occurs in about 3% of untreated patients, may be accompanied by tenosynovitis, particularly at the wrists or in the dorsa of the hands. The arthritis can be of two types. First is the painful, swollen joint which, on tapping, reveals purulent exudate under pressure and is associated with Gram-negative intracellular diplococci. The culture is usually positive, and the patient normally has symptoms and signs of gonorrhoea. The second type is a painful joint with little or no swelling and no increased heat of the local skin. On aspiration the fluid is not under pressure, contains few cells and gives a negative culture for gonococci or other organisms. The patient may lack subjective symptoms of gonorrhoea but gonococci can be cultured from the urogenital tract. This type of arthritis frequently occurs together with gonococcal dermatitis.

Gonococcaemia may produce a typical clinical picture with almost pathognomonic skin lesions. Vesicular, pustular or haemorrhagic lesions develop on erythematous patches of the skin, usually localized to the extensor surfaces of the extremities.

The cutaneous lesions usually develop in connection with septic fever and arthralgia. During the fever stages a blood culture for gonococci is frequently positive. Meningitis and endocarditis are rarely seen, though they may occur when the gonococcus invades the blood stream. Gonococcaemia occurs especially in cases with long-standing infection of asymptomatic nature, e.g., in females and homosexuals with proctitis.

Gonococcal perihepatitis, which is almost without exception due to intraperitoneal spread of gonococci, is probably a much more frequent complication than reported.

ASYMPTOMATIC GONORRHOEA

Asymptomatic gonorrhoea is recognized as one of the most important causes in the continuing rise in incidence of this disease. By definition, it is the stage in which no clinical signs or symptoms of the disease are detected either by the patient or by the examining physician. In this asymptomatic category those patients with signs or symptoms which can mimic those of gonorrhoea but which are proved to be caused by other diseases may also be included. Essentially, the asymptomatic state is diagnosed solely on the basis of the laboratory tests.

Unfortunately, even the best culture methods are probably not sensitive enough to diagnose all cases of asymptomatic infection. In most studies on female contacts of infected males about 50–70% of these females show positive cultures. Undoubtedly some of the culture-negative patients have gonorrhoea. Therefore, studies on routine examination by culture will not reveal the total number of infections.

For a number of years asymptomatic gonorrhoea was thought to be confined primarily to the female genital tract. More intensive study of the clinical aspects of the disease have shown that the rectum and the pharynx, as well as the male urethra, may also harbour the organism asymptomatically. The question naturally arises as to whether these asymptomatic patients are harbouring a nonpathogenic strain of gonococcus. This is not the case, since the gonococcus, while capable of existing in these sites without producing symptoms, can later produce serious complications in some of these carriers and, when transmitted to others, may produce a symptomatic infection in the contact.

In a growing number of countries the majority of women found to have gonorrhoea are those named as contacts of the disease. Between 70 and 80% are asymptomatic.

It is a feature of the modern western world that more and more asymptomatic women seek a check-up or ask outright for tests for sexually transmitted disease. Men have been doing this for decades. These socially legitimate demands have increased with growing sexual freedom and, no doubt, with public education. Other women present themselves with genito-urinary symptoms which are unsupported by the clinical findings. In such cases it is nevertheless up to the physician to be highly suspicious of the possibility of gonorrhoea and to repeat tests in order to achieve an optimal level of positive cultures.

The phenomenon of the male carrier has recently received new emphasis. Thoroughly emancipated young women in the present age of sexual freedom are more frank and honest about their sexual activities than their mothers were 20–30 years ago. Hence their contacts present for examination more often and the asymptomatic male carrier is, therefore, seen more frequently, now constituting 10–20% of all male cases with gonorrhoea.

The asymptomatic male may or may not have signs of the infection. The percentage found with signs will be increased if early morning examination is made. This applies particularly to men named as sex contacts of women with proven gonorrhoea. It applies also to men who have submitted themselves to self or quack treatment, as so often happens in many parts of the world. In these conditions under-dosage, either with a sulfonamide or with an antibiotic, seems to be usual. In these cases there may, in the morning, be evidence of urethral discharge or the external meatus may be found lightly sealed with a thin crust of dried secretion.

Asymptomatic carriers of gonococci are of great importance in maintaining gonorrhoeal infections at constantly high levels. One of the main tasks in preventing the spread of gonococcal infections is the tracing and treating of these cases, often called the gonococci reservoir.

ANTIBIOTIC RESISTANCE IN *NEISSERIA GONORRHOEAE*

Apparent lack of response to therapy may be due to reinfection, persistence of concomitant nongonococcal infection or true treatment failure. Reasons cited for failure of antibiotics to cure gonococcal infection have been diverse. However, the primary cause of treatment failure in most studies has been increased resistance of the gonococcus to antibiotics. Penicillin is still the most commonly used drug in the treatment of gonorrhoea. The reports of strains of *Neisseria gonorrhoeae* producing penicillinase since 1976 have, therefore, been of the utmost clinical and epidemiological importance. Stolz and co-workers [4] reported recently that in 1979 β-lactamase-producing gonococci represented 4% of strains isolated in the Rotterdam area. In 1978 the corresponding figure in Finland was 3% [5]. The β-lactamase-producing strains have spread from the Far East but now it seems, at least as indicated by Dutch and Finnish findings, they are gradually becoming established in Europe, also. This further stresses the importance of culture, identification and sensitivity testing of gonococci. It may well also mean that the routine treatment of gonorrhoea with penicillins will have to be changed and another antibiotic used.

Relative resistance to penicillin has gradually increased during the last two decades. The proportion of relatively penicillin-resistant strains seems to vary between different parts of the world. Gonococcal strains isolated in or derived from the Far East seem exceptionally resistant to penicillin. Gonococci which are relatively resistant to penicillin often show cross-resistance to other antibiotics. In a Finnish survey such strains do not so far seem to have any greater therapeutic significance.

THE NEED FOR IDENTIFICATION OF *NEISSERIA GONORRHOEAE*

A diagnosis of gonorrhoea cannot be made on symptoms or clinical signs only but must be based on the demonstration of gonococcal organisms. Examination of a Gram-stained smear taken directly from the patient is a rapid and fairly accurate diagnostic method in symptomatic men. Experience shows that a rapid presumptive diagnosis can be arrived at by direct microscopy in approximately 85–95% of males with acute gonorrhoea. However, the diagnostic yield with this technique is much lower in females and dependent to a great extent upon the person behind the microscope. Thus in several investigations during recent years culture procedures have been shown to be superior and the yield with direct microscopy as compared to culture has varied from 20 to 80%. Corresponding difficulties are also encountered in males with asymptomatic gonorrhoea as well as in patients with complicated gonorrhoea. The high incidence of asymptomatic gonorrhoea makes culture procedures increasingly important since the use of direct microscopy alone would miss a large proportion of these cases. Gonococcal cultures together with appropriate methods of identification are of special importance for detecting asymptomatic infections in the rectum and pharynx. Only an early and correct diagnosis of symptomatic or asymptomatic gonorrhoea combined with adequate treatment prevents with certainty the development of acute or chronic complications.

The decreasing sensitivity of gonococci to antibiotics, especially penicillin, also makes culture of gonococci essential. Today it is of the utmost importance to trace all new infections by β-lactamase-producing gonococci to prevent their further spread. If this is not done properly these gonococci will, sooner or later, necessitate a change in routine treatment from penicillin to some much more expensive and probably less safe antibiotic. Furthermore, complications caused by β-lactamase-producing strains are definitely more serious.

It may be concluded that experience in many countries, both developing and developed, shows clearly that sound, scientific diagnosis of gonorrhoea is not only in the best interest of individual patients but is the only solid foundation for epidemiological control and prevention of disability. The vast amounts of money and time spent on bed occupancy and operations for acute and chronic pelvic complications in women has been recently highlighted by Rendtorff and co-workers [6] in the United States of America. The situation is worse in many underdeveloped countries. In world-wide terms the cost of gonorrhoea is enormous and rising. Provision of a culture service including identification and sensitivity testing therefore makes sound economic sense in any language.

REFERENCES

1. Bro-Jørgensen, A. and Jensen, T. (1973): Gonococcal pharyngeal infections. Report of 110 cases. *Br. J. Vener. Dis.*, *49*, 491.
2. Ödegaard, K. and Gundersen, T. (1973): Gonococcal pharyngeal infections. *Br. J. Vener. Dis.*, *49*, 350.
3. Willmott, F.E. (1974): Transfer of gonococcal pharyngitis by kissing? *Br. J. Vener. Dis.*, *50*, 317.
4. Stolz, E., Nayyar, K.C. and Noble, R. Unpublished data.
5. Jakhola, M. and Lassus, A. Unpublished data.
6. Rendtorff, R.C., Curran, J.W., Chandler, R.W., Wiser, W.L. and Robinson, H. (1974): Economic consequences of gonorrhoea in women. Experience from an urban hospital. *J. Vener. Dis. Assoc.*, *1*, 40.

Discussion

I. Lind (Copenhagen, Denmark): Has Professor Lassus any figures for the incidence of complications in Finland?

A. Lassus: I only have figures from my own hospital. We have about 2,500 new cases of gonorrhoea each year. The frequency of complications in males is about 3–4% prostatitis and epididymitis (these are still the most common complications) and about 10–15 new cases of gonococcaemia per year. About 25% of the patients with acute salpingitis in the Helsinki area have gonorrhoea and about 30% have a chlamydial infection. In the Helsinki area the incidence of pharyngeal gonorrhoea is about 6%.

A. J. M. van Alphen (Amsterdam, The Netherlands): If triple tests are done in asymptomatic patients, what is the time interval between the tests carried out? Are all the tests carried out in one day or over several days?

A. Lassus: You are referring to the re-testing of female patients?

A. J. M. van Alphen: Yes.

A. Lassus: When the patients returned after one week and were found to be negative they were re-tested and tested again after another week. Of course, there is a risk of reinfection in such a series.

E. Randall (Evanston, U.S.A.): Recently, at a meeting in Chicago, it was reported that chlamydia rather than gonococcus had been isolated from about seven people with symptoms of perihepatitis. I wonder whether what we have been calling perihepatitis due to gonococcus, simply because the gonococcus is found in the cervix, might not be due to chlamydia. Has Professor Lassus found any cases associated with chlamydia?

A. Lassus: I have seen both chlamydial and gonococcal cases. I reported one case (*Br. J. Vener. Dis.* 1973, *49*, 83) with gonococcal sepsis, perihepatitis and a positive gonococcal finding in the cervix. We did not search for chlamydia in that case.

J. Wallin (Uppsala, Sweden): Another point which could be added to the list of reasons for the need for identification of *Neisseria gonorrhoeae* concerns countries where there is a legal aspect to a diagnosis of gonorrhoea. This means that we have, perhaps, to be more careful than in many other situations to be absolutely sure that the diagnosis of gonorrhoea is correct.

A. Lassus: The law was changed in Finland about two years ago. It is no longer possible to obtain a divorce in that country because of gonorrhoea.

G. L. Daguet (Paris, France): I would like to mention an unusual location of gonorrhoea in a newborn. We recently cultivated gonococcus from gastric juice aspirate in a newborn whose mother had asymptomatic gonorrhoea. Has anyone else made such an observation?

A. Lassus: I have never heard of anything like that.

J. Maeland (Trondheim, Norway): Two days before coming to this meeting we isolated gonococci from the gastric aspirate of a newborn child.

Immunological aspects of
Neisseria gonorrhoeae

E. Sandström and D. Danielsson
Departments of Dermatology and Clinical Bacteriology, Karolinska Institute, Southern Hospital, Stockholm and Department of Clinical Bacteriology and Immunology, Central County Hospital, Örebro, Sweden

INTRODUCTION

Over the last decade our knowledge of the gonococcal organism has increased at an accelerating rate. We are approaching a state in which these are among the best known Gram-negative organisms.

Both extracellular and intracellular [1, 2] components have been utilized in serological work but most studies have dealt with more or less well-defined surface components. The picture of the gonococcal surface that is now emerging is similar to that of other Gram-negative organisms.

On the gonococcal outer membrane the lattice of lipopolysaccharide and major outer membrane protein (MOMP) is accessible to serological studies. The components are difficult to separate and are therefore often studied as a complex. Other components such as pili, acid polysaccharides and 'minor' proteins have been identified and found immunogenic. Lately, capsular substances have been observed by several investigators. For review and assessment of the current state of knowledge, two excellent collections of papers are now available [3, 4].

Several of these substances have been utilized in efforts to construct a much-needed serological classification of *Neisseria gonorrhoeae* (Table I) using a variety of techniques. (For a review see [5].)

The older work, primarily using agglutination, supported the notion of two major groups with intermediate forms [6]. Some strains were, however, found to be antigenically 'broader' than others [7]. A great step forward was taken with the thorough agglutination studies by Wilson [8]. On whole cells heated to 100°C in suspension he could identify three group antigens, A, B_1 and B_2, one or more of which was always present. A fourth antigen, C, also behaved as a group antigen. Thus each strain could be identified by a formula. The results could be confirmed in the complement fixation test with a 'protein' antigen. He found, however, that the group antigens could be gained or lost on subculture, while it was always the same type antigen that was lost or regained in a given strain. This made the already laborious antiserum production and absorptions even more cumbersome. Unfortunately, work along these promising lines was not pursued. In fact, most work on gonococcal serology seems to have been put off for almost a decade until Maeland presented a classification based on a periodate-sensitive and often lactose-inhibitable

22

Table I. Published classification studies of Neisseria gonorrhoeae.

Author [Reference]	Year	Method	Antigen	Results
Wilson and Miles [6]	1949	Summary, mostly agglutination with whole cells		Two major groups with intermediate forms
Reyn [27]	1949	Complement fixation	Whole cells	Five major groups
Wilson [8]	1954	Agglutination (complement fixation)	Whole cells (protein extracts)	Group antigens A, B_1, B_2 and C, alone or in combination. Type antigens D, E, F and G. Antigenic formula. Changes on subculture.
Danielsson [28]	1965	Immunofluorescence	Whole cells	Four major and some minor groups
Maeland [11]	1969	Haemagglutination & passive haemolysis	Endotoxin	Six factors. Antigenic formula. Periodate sensitive. Some inhibited by lactose
Glynn and Ward [29]	1969	Normal human and hyperimmune serum bactericidal	Live whole cells	Four overlapping groups
Geizer [30]	1975	Gel diffusion	Heat extract	Five serotypes
Apicella [12]	1976	Gel diffusion & haemagglutination inhibition	Acid polysaccharide	Gc_1, Gc_2, Gc_3, Gc_4 (and Gc_5). Some strains with more than one type
Johnston *et al.* [13]	1976	Gel diffusion	Major outer membrane protein complex	Sixteen serotypes
Tramont [19]	1976	Anti-*N. meningitidis* serum bactericidal	Live whole cells	Eight absorbed sera (20 reagents). Strain specific formula
Wong *et al.* [25]	1976	Immunotyping (serum bactericidal, agglutination)	Live whole cells	Discussion of correlation
Sidberry *et al.* [31]	1977	Pyocin	Live whole cells	Thirty patterns with 23 extracts
Wang *et al.* [15]	1977	Micro-immunofluorescence	Whole cells	Three nonoverlapping groups with subgroups. Correlation with disseminated infection
Brinton *et al.* [32]	1978	ELISA	Pili	At least 29 types. Some pili of more than one type
Buchanan *et al.* [33]	1978	ELISA	Principal outer membrane protein	At least six serotypes. Some strains with more than one serotype

determinant in the endotoxin [9, 10]. He was able to identify six determinants in various combinations, using three strains and the same logic as Wilson [11]. This work has been extended with more purified acid polysaccharides by Apicella [12], although the identity of the respective antigens is not proven. To date, five serotypes have been identified that are not shared by the two absorbing strains. However, some of the tested strains (12%) were found to have more than one serotype [12]. Using antigens purified with regard to the MOMP, Johnston and co-workers [13] were able to identify 16 serotypes. The protein was, however, in a large complex ($> 10^6$ daltons) and has since been found to be dissociable with non-ionic detergents to yield cross-reactivity between hitherto distinct serotypes [14]. This work greatly stimulated interest in a gonococcal classification. Studies along more traditional lines by Wang and co-workers [15] have yielded the most promising system so far. Using the micro-immunofluorescence (micro-IF) method, with whole cells as antigen, three non-overlapping groups with subgroups could be identified using mouse antisera absorbed with a single 'unrelated' broadly-reacting strain. Group A could be correlated with the auxotype arginine$^-$, hypoxanthine$^-$, uracil$^-$ and isolation from patients with disseminated gonococcal infection. The performance of the micro-IF method, however, requires special skills, which has hindered its wider spread.

THE DEVELOPMENT OF A NEW CLASSIFICATION SYSTEM

When the co-agglutination (COA) technique was adapted to the identification of the gonococcus in the laboratory [16], a simple, rapid and economical method was potentially available for gonococcal classification. It could be shown that results in the COA correlated with those obtained with the immunofluorescence and agar gel precipitation [17] methods used in most earlier typing studies. Analysis by immunization, absorption and agglutination was found to be as laborious and time-consuming as reported by Wilson [8], since the evidence was that more than one antigen was reactive in the COA, too.

Utilizing the correlation with agar gel precipitation the line-rocket immuno-electrophoresis (L-RIE) method was developed as a method for analysing the relationships between different antigens in a complex system [18]. Basically, the test antigens are moulded into an intermediate gel and their rockets identified by correspondence with the reaction lines formed by a reference antigen and antiserum system. The analysis of such an experiment is shown in Table II. Cx signifies an identity between an antigen in D-6 and V-15, while cr describes an antigen (i.e., a rocket) in common with the test strain and D-6 but without information on the relationship to other test antigens. The COA results when anti-D-6 is absorbed with the strains reactive in the L-RIE are shown beneath.

Three different patterns are seen after absorptions with B-2, C-3 and B-2 & C-3. With the aid of the pattern obtained after absorption of C-3, the MOMP serotype strains could be divided in two groups and the antigen that was responsible for this reaction was tentatively called W. After absorption with B-2 another cross-reacting antigen, M, could be observed and after absorption with both of these strains an antigen J was observed that correlated with the cx reaction in the L-RIE. This kind of analysis was carried out for all the MOMP serotype reference strains in a search for as many antigen patterns as possible. This resulted in a panel of absorbed anti-

24

Table II. L-RIE and COA results with anti-MOMP reference strain D-4 antiserum.

| | L-RIE | | | | | | | | | | | | | | | |
---	W	B	G	E	R	D	V	N	H	C	X	S	T	U	A	F
	nd	–	–	–	–	cx	cx	–	–	–	–	–	–	–	–	–
	nd	cr	cr	cr	cr	cr	cr	–	–	cr	–	–	–	–	–	–
COA reag. abs. with																
W-16	–	3+	2+	3+	3+	2+	3+	2+	2+	3+	–	–	–	–	+	2+
B-2	–	–	–	–	3+	3+	3+	–	–	3+	–	–	–	–	–	–
G-7	+	+	–	2+	2+	3+	3+	–	–	3+	–	–	–	–	–	–
E-5	–	–	–	–	+	3+	2+	–	–	–	–	–	–	–	–	–
R-11	–	–	–	–	–	+	–	–	–	–	–	–	–	–	–	–
D-4	–	–	–	–	–	–	–	–	–	–	–	–	–	–	–	–
V-15	–	–	–	–	+	+	–	–	–	–	–	–	–	–	–	–
C-3	2+	+	+	+	3+	3+	3+	–	–	–	–	–	–	–	–	–
A-1	–	2+	–	–	3+	3+	3+	–	–	2+	–	–	–	–	–	–
F-6	nd	3+	2+	3+	3+	3+	3+	+	+	3+	–	–	–	2+	–	–
B-2+C-3	–	–	–	–	–	3+	3+	–	–	–	–	–	–	–	–	–

cx = reaction of identity with line in reference system (D-6 vs anti-D-6)
cr = rocket above test antigen, relationship to other rockets not shown
nd = not done

sera. Seven were used for detection of W antigens, seven for M antigens and five for J antigens. Using this panel, the different antigen classes were further characterized.

THE WMJ-SYSTEM

Class W antigens

The MOMP serotype strains were found to be divided into three non-overlapping groups with class W reagents (Table III), a situation similar to that in the micro-IF

Table III. COA patterns with reagents against the tentative antigen class W and the MOMP reference strains.

| | Group | | | | | | | | | | | | | | | |
---	W	B	G	E	R	D	V	N	H	C	X	S	T	U	A	F
	–	3+	3+	3+	3+	1+	3+	–	–	–	–	–	–	–	–	–
W/I	2+	1+	1+	1+	3+	3+	3+	–	–	–	–	–	–	–	–	–
	–	–	–	–	2+	2+	3+	–	–	–	–	–	–	–	–	–
	–	–	–	–	–	–	–	3+	3+	3+	3+	3+	–	–	–	–
W/II	2+	–	–	–	–	–	–	3+	3+	3+	3+	3+	3+	3+	–	–
	–	–	–	–	–	–	–	2+	2+	2+	2+	–	2+	2+	–	–
W/III	–	–	–	–	–	–	–	–	–	–	–	–	–	–	2+	3+

system [19]. Strain W-16 reacted with one reagent specific for each of the groups W/I and W/II but the antiserum used for the W/II reagent was found to contain antibodies to a class M antigen in strain W-16. The strain also gave rise to antibodies specific for W/I and was thus assigned to this class. When the reference strains for the micro-IF method were tested, groups B and C could not be clearly separated. However, the MOMP serotype strains were classified analogously in the micro-IF method except for strain A-1 which was classified in group B (Wang, personal communication). The immunotyping strains were classified analogously in both systems, supporting the correspondence of COA groups W/I, W/II and W/III to micro-IF groups A, B and C respectively [20]. Strains from patients with disseminated gonococcal infections were classified in group W/I except for one strain which was placed in group W/II and which was also reported to belong to micro-IF group B [20]. All tested strains of the arginine$^-$, hypoxanthine$^-$, uracil$^-$ auxotype $(n = 11)$ belonged to group W/I [5]. No cross-reactions between meningococci and the class W reagents have been observed [5]. Fifty-eight unselected clinical isolates could all be classified regardless of colony morphology variant [21]. Thirty-four of these strains were from 14 patients and all strains from the same patient belonged to the same group. In this series another absorbed antiserum appeared a better representative for W/III, as already suggested by the results with micro-IF method reference strains, in order to avoid cross-reaction between W/II and W/III [21].

Class M antigens

The class M antigens were found to be sensitive to oxidation by periodate, as reported for the endotoxin factors [9]. When the reference strains for that system were tested (Table IV), patterns corresponding to factors a_2, a_3, a_5 and a_6 were found. It was also found that sugar inhibitions could be performed in the COA and that lactose inhibited factors a_5 and a_6 but not a_3, in accordance with published results [10]. When the same tests were carried out with the reference strains for the acid poly-

Table IV. COA patterns with reagent against the tentative antigen class M and the endotoxin reference strains. Effect of pronase and periodate. Blocking by lactose.

Factor	8551			V			VII			Endotoxin fact. pattern
	pro[a]	per[b]	sug[c]	pro	per	sug	pro	per	sug	
M/a	res	sens	lact	res	sens	lact	–	–	–	a_6
M/b	res	sens	lact	–	–	–	–	–	–	a_5
M/c	res	sens	lact	–	–	–	–	–	–	a_5
M/d	–	–	–	res	sens	NB	res	sens	NB	a_3
M/e	–	–	–	res	sens	nd	–	–	–	a_2
M/f	–	–	–	–	–	–	–	–	–	–
M/g	–	–	–	–	–	–	–	–	–	–

[a] pro = pronase, [b] per = periodate; res = resistance, sens = sensitive, [c] sug = blocking by sugar; lact = blocking by lactose; NB = no blocking by lactose, glucose, galactose, glucosamine; nd = not done

saccharide system [12], it was found that some of the absorbed antisera considered specific for class M contained reactivity against pronase-sensitive antigens. Gc_4 corresponded to strain 8551 and Gc_2 to strains V and VII, as reported [12]. All acid polysaccharide reference strains were found to contain more than one M factor.

With 34 strains from 14 patients it was found that eight patients had strains with different class M patterns in different anatomical sites, in contrast to the finding with class W referred to above [21]. This difference did not correlate with colony morphology.

Class J antigens

These antigens probably represent a heterogeneous group. They were originally selected with the aim of identifying the MOMP serotype antigens. The MOMP reference strains were used in the immunizations and absorptions for all the reagents described above but only two reagents have been found specific for a single MOMP reference strain, in spite of persistent search. One of these even turned out to be better characterized as a group W/III reagent [21]. Two of the remaining four reagents reacted with antigens sensitive to pronase. Twenty-five of 58 unselected clinical isolates reacted with these four reagents. Five of these (20%), however, reacted only with the opaque colony variant but not with the transparent variant [22, 23]. In a parallel series of experiments it was observed that opaque colonies gave rise to 'strain-specific' antibodies in immunizations and absorbed and agglutinated these antibodies but that the transparent colony morphology variant did not [21]. These reactive antigens were sensitive to pronase but not to periodate.

Tests of the practical utility of the WMJ system

A serological system is not an end in itself but a tool to serve many purposes. With the COA method the WMJ system has been tested in a few applications.

Some strains presumed to be identical have been obtained from different sources in the course of these studies [24] (Table V). A good correspondence was observed between strains 8551 and Gc_4 and between NRL 6218 and C-33. The situation is, however, more complex with three strains supposed to be F62: KF [25], NRL 5293 [15] and F62 (Buchanan, personal communication). The two latter differ from the first and strongly resemble the 2686 from Atlanta [25] and Texas (A-1 [13]), suggesting that some strain(s) might be misnamed. The same difference has been observed in the micro-IF for KF and NRL 5293 [20].

The results with 46 strains from six patients with β-lactamase-producing gonococci have been reported in detail [24]. The general conclusions were that class W antigens remained stable for more than two months in the throat of one patient and was the same in two partners. This was true for class J too. Class M antigens were, however, seen to change in repeated isolates from two patients, which might reflect a mixed infection or selection of new variants. The clinical usefulness has since been confirmed in three new cases with β-lactamase-producing gonococci that were diagnosed within a two-week period. Strains from two of the patients were serologically almost identical and clearly different from the third. These two patients could subsequently be firmly linked without possibility of contact with the third who, on intensive questioning, yielded a new chain of infection with a culture-negative intermediate.

Table V. COA patterns with reagents against tentative classes W, J and M and presumably identical strains. (For abbreviations, see Table IV.)

		8551	GC$_4$	6218	C-33	KF	5293	F62	2686	A-1
Group	W/I	–	–	–	–	–	–	–	–	–
		–	–	–	–	–	–	–	–	–
		–	–	–	–	–	–	–	–	–
		–	–	–	–	3+	–	–	–	–
	W/II	–	–	3+	3+	3+	–	–	–	–
		2+	2+	3+	3+	–	3+	2+	3+	–
	W/III	3+	3+	–	–	–	3+	3+	3+	2+
Type	J/4	–	–	–	–	–	–	–	–	–
	J/14	–	3+	3+	3+	–	1+	1+	3+	–
	J/9	–	–	–	–	–	–	–	–	–
	J/12	–	–	–	–	3+	–	–	–	–
	J/6	3+	3+	–	–	–	–	–	–	–
Factor	M/a	3+	3+	–	–	2+	–	–	–	–
	M/b	3+	3+	–	–	–	1+	1+	–	–
	M/c	3+	3+	–	–	–	–	–	–	–
	M/d	–	–	3+	2+	3+	3+	3+	3+	3+
	M/e	–	–	1+	–	3+	2+	1+	2+	–
	M/f	–	–	–	–	–	–	–	–	–
	M/g	–	–	–	–	–	3+	1+	3+	2+

CONCLUSIONS

The WMJ classification in the COA method appears interesting from several points of view.

COA is a simple, rapid and economical method that can be widely disseminated into research as well as into clinical laboratories. Results correlate strongly with results from immunofluorescence and from agar gel precipitation of surface and soluble antigens [17, 21]. The method allows the study of pronase digestion and periodate oxidation of surface antigens and is sensitive to blocking of reactive groups with, for example, simple sugars. It offers a tool for identification of strains for antigen production for serological tests and for interlaboratory standardization of reference strains.

With the observation of the changing class M pattern there must be concern about the R-type lipopolysaccharides proposed as a vaccine [26]. The WMJ system could be used in the production of antigens for immunizations and the necessary serological tests and for the evaluation of the challenges.

Serological epidemiology would be possible on three levels. Firstly, on a geographical and time scale in order to monitor the patterns of gonococcal spread in connection with the search for points of attack accessible to legal or policy measures, secondly in the effort to identify strains that carry higher risks of complications of, for example, disseminated gonococcal infection and thirdly, on the clinical level, to trace particular chains of infection, for example of β-lactamase-producing gonococci with non-β-lactamase- or culture-negative links, or to distinguish between relapses and reinfections in therapeutic trials.

Class W seems to be of great potential in serving as an epidemiological marker since it appears stable and clinically relevant. Subgrouping seems to be feasible. Class M may be of primary importance in the study of the host immune defence and class J promises to be a simple way of identifying surface proteins important for the understanding of the infectious process.

SUMMARY

A new approach to the analysis of complex antigen systems is described in the light of current knowledge of gonococcal classification systems. Three classes of antigens active in the co-agglutination (COA) test are found on the gonococcal surface. These antigens vary independently of one another, giving each strain its own 'fingerprint'. The first class, W, is found to correlate closely with the micro-immunofluorescence typing system. The second, M, shares periodate sensitivity and lactose inhibition with the endotoxin system. The third class, J, is probably heterogeneous. Colony morphology-dependent, pronase-sensitive antigens are found in this class.

The implications of the COA classification of *Neisseria gonorrhoeae* presented are discussed.

ACKNOWLEDGEMENTS

This work was supported by grants from the Karolinska Institute, the Swedish Medical Research Council and the World Health Organization.

REFERENCES

1. Kwapinski, G., Kwapinski, E. and Webb, C.J. (1978): Studies on circulating gonococcal antibodies and antigens. *Can. J. Microbiol.*, 24, 109.
2. Schmale, J.D., Danielsson, D.G., Smith, J.F., Lee, L. and Peacock, W.L. (1969): Isolation of an antigen of *Neisseria gonorrhoeae* involved in human immune response to gonococcal infection. *J. Bacteriol.*, 99, 469.
3. Brooks, G.F., Gotschlich, E.C., Holmes, K.K., Sawyer, W.D. and Young, F.E. (Editors) (1978): *Immunobiology of Neisseria gonorrhoeae*. American Society for Microbiology, Washington, D.C.
4. Roberts, R.B. (Editor) (1977): *The Gonococcus*. John Wiley & Sons, Inc., New York.
5. Sandström, E. (1979): Studies on the serology of *Neisseria gonorrhoeae*. Thesis, Stockholm.
6. Wilson, G.S. and Miles, A.A. (1946): Antigenic structure of the gonococcus. In: *Topley and Wilson's Principles of Bacteriology and Immunity*, 3rd ed., p. 543. E.J. Arnold, London.
7. Torrey, J.C. (1940): A comparative study of antigens for the gonococcal complement fixation test. *J. Immunol.*, 38, 413.
8. Wilson, J.F. (1954): A serological study of *Neisseria gonorrhoeae*. *J. Pathol. Bacteriol.*, 68, 495.
9. Maeland, J.A. (1968): Antigenic properties of various preparations of *Neisseria gonorrhoeae* endotoxin. *Acta Pathol. Microbiol. Scand.*, 73, 413.
10. Maeland, J.A., Kristoffersen, A. and Hofstad, T. (1971): Immunochemical investigations on *Neisseria gonorrhoeae* endotoxin. *Acta Pathol. Microbiol. Scand.*, 79, 233.

11. Maeland, J.A. (1969): Serological cross-reactions of aqueous ether extracted endotoxin from *Neisseria gonorrhoeae* strains. *Acta Pathol. Microbiol. Scand.*, *77*, 505.

12. Apicella, M.A. (1976): Serogrouping of *Neisseria gonorrhoeae*: identification of four immunologically distinct acidic polysaccharides. *J. Infect. Dis.*, *134*, 377.

13. Johnston, K.H., Holmes, K.K. and Gotschlich, E.C. (1976): The serological classification of *Neisseria gonorrhoeae. J. Exp. Med.*, *143*, 741.

14. Johnston, K.H. (1977): Surface antigens: an outer membrane protein responsible for imparting serological specificity to *Neisseria gonorrhoeae*. In: *The Gonococcus*, p. 273. Editor: R.B. Roberts. John Wiley & Sons, Inc., New York.

15. Wang, S.P., Holmes, K.K., Knapp, J.S., Ott, S. and Kyzer, D.D. (1977): Immunological classification of *Neisseria gonorrhoeae* with microimmunofluorescence. *J. Immunol.*, *119*, 795.

16. Danielsson, D. and Kronvall, G. (1974): Slide agglutination method for the serological identification of *Neisseria gonorrhoeae* with antigonococcal antibodies absorbed to protein A-containing staphylococci. *Appl. Microbiol.*, *27*, 368.

17. Danielsson, D. and Sandström, E. (1979): Serology of *Neisseria gonorrhoeae*. Demonstration of strain-specific antigens with immunoelectrophoresis, immunofluorescence and co-agglutination techniques. *Acta Pathol. Microbiol. Scand. Sect. B*, *87*, 55.

18. Sandström, E. and Danielsson, D. (1979): Serology of *Neisseria gonorrhoeae*: characterization of hyperimmune antisera by line-rocket immunoelectrophoresis for use in co-agglutination. *Acta Pathol. Microbiol. Scand. Sect. B*. To be published.

19. Tramont, E.C., Griffis, J.M., Rose, D., Brooks, G.F. and Artenstein, M.S. (1976): Clinical correlation of strain differentiation of *Neisseria gonorrhoeae. J. Infect. Dis.*, *134*, 128.

20. Mark, J.A. and Wang, S.P. (1978): Comparison of antigenic heterogeneity of *Neisseria gonorrhoeae* strains by microimmunofluorescence and serum bactericidal tests. *Infect. Immun.*, *22*, 403.

21. Danielsson, D. and Sandström, E. (1979): Serology of *Neisseria gonorrhoeae*: Demonstration with co-agglutination and immunoelectrophoresis of antigenic differences associated with colour/opacity colonial variants. *Acta Pathol. Microbiol. Scand. Sect. B*. To be published.

22. Swanson, J. (1978): Studies on gonococcus infection. XIV. Cell wall protein differences among colour/opacity colony variants of *Neisseria gonorrhoeae. Infect. Immun.*, *21*, 292.

23. Walstad, L.D., Guymon, L.F. and Sparling, P.F. (1977): Altered outer membrane protein in different colonial types of *Neisseria gonorrhoeae. J. Bacteriol.*, *129*, 1623.

24. Sandström, E. and Danielsson, D. (1979): Serology of *Neisseria gonorrhoeae*: classification with co-agglutination. *Acta Pathol. Microbiol. Scand. Sect. B*. To be published.

25. Wong, K.H., Arko, R.J., Logan, L.C. and Bullard, J.C. (1976): Immunological and serological diversity of *Neisseria gonorrhoeae*: gonococcal serotypes and their relationship with immunotypes. *Infect. Immun.*, *14*, 1297.

26. Diena, B., Ashton, F.E., Wallace, A.R.A. and Perry, M.B. (1978): Immunizing properties of gonococcal (R-type) lipopolysaccharide in the chicken embryo animal model. In: *Immunobiology of Neisseria gonorrhoeae*, p. 319. Editors: G.F. Brooks, E.C. Gotschlich, K. K. Holmes, W.D. Sawyer and F.E. Young. American Society for Microbiology, Washington, D.C.

27. Reyn, A. (1949): Serological studies on gonococci. II. Cross-absorption experiments and factor serum determinations. *Acta Pathol. Microbiol. Scand.*, *26*, 234.

28. Danielsson, D. (1965): The demonstration of *Neisseria gonorrhoeae* with the aid of fluorescent antibodies. 3. Studies by immunofluorescence and double diffusion-in-gel technique on the antigenic relationship between strains of *N. gonorrhoeae. Acta Pathol. Microbiol. Scand.*, *64*, 243–266.

29. Glynn, A.A. and Ward, M.E. (1970): Nature and heterogeneity of the antigens of *Neisseria gonorrhoeae* involved in the serum bactericidal action. *Infect. Immun.*, *2*, 162.

30. Geizer, I. (1975): Studies on serotyping of *Neisseria gonorrhoeae. Zentralbl. Bakt. Parasitenkd. Infektionskr. Hyg. Abt. 1 Orig. A*, *232*, 213.

31. Sidberry, H. and Sadoff, J.C. (1977): Pyocin sensitivity of *Neisseria gonorrhoeae* and its feasibility as an epidemiological tool. *Infect. Immun., 15,* 628.

32. Brinton, C.C., Bryan, J., Dillon, F.A., Guerina, N., Jacobson, L.J., Labic, A., Lee, S., Levine, A., Lim, S., McMichael, J., Polen, S., Rogers, K., To, A.C.C. and To, S.C.M. (1978): Uses of pili in gonorrhoea control. In: *Immunobiology of Neisseria gonorrhoeae,* p. 155. Editors: G.F. Brooks, E.C. Gotschlich, K.K. Holmes, W.D. Sawyer and F.E. Young. American Society for Microbiology, Washington, D.C.

33. Buchanan, T.M., Chen, K.C.S., Jones, R.B., Hildebrandt, J.F., Pearce, W.A., Hermodson, M.A., Newland, J.C. and Luchtel, D.L. (1978): Pili and principal outer membrane protein of *Neisseria gonorrhoeae*: immunochemical, structural and pathogenic aspects. In: *Immunobiology of Neisseria gonorrhoeae,* p. 145. Editors: G.F. Brooks, E.C. Gotschlich, K.K. Holmes, W.D. Sawyer and F.E. Young. American Society for Microbiology, Washington, D.C.

Laboratory media for isolation and specimen transport in the United States

J. E. Martin and J. S. Lewis
Department of Health, Education and Welfare, Public Health Service, Center for Disease Control, Bureau of Laboratories, Atlanta, U.S.A.

At the invitation of the League of Nations, representatives from countries of the western world met during the late 1920s to update their knowledge of the venereal diseases. As a result of some of the reports presented at this meeting, several foundations and wealthy individuals in the United States sponsored research programmes towards the ultimate control of these diseases. Through the efforts of Dr John Parran, Surgeon General of the United States Public Health Service in the early 1930s, federal funds were made available to support the initial federal programmes. Research interests were directed primarily towards syphilis while in the area of gonorrhoea efforts were made to implement the diagnosis and treatment programmes developed in Europe.

The availability of continued funding, coupled with the involvement of the United States in World War II, created an interest among several experts in the isolation and identification of the gonococcus. By 1945 this interest had grown and produced a multitude of special culture media used in the laboratory diagnosis of gonorrhoea.

In most instances, according to the investigator, each medium was superior to those previously recommended. Comparative studies by other workers not infrequently failed to confirm the original observations. In order to determine the most suitable medium for use in public health and clinical laboratories a group of bacteriologists experienced in the cultivation of the gonococcus volunteered to undertake an evaluation of the 12 media most commonly used, under uniform conditions, in one laboratory where cultures of the same specimen could be made on all 12 media [1]. The results of that study revealed that the media detecting the highest number out of a total of 70 strains were:

1) a modified McLeod's agar with Nile Blue A, enriched with horse-plasma and haemoglobin,

2) GC agar base, with Bacto-haemoglobin and Supplement B,

3) Proteose no. 3 agar with Nile Blue A, enriched with horse-plasma and haemoglobin.

These media recovered 65, 63 and 63 strains, respectively. The GC base agar was chosen as the medium to be recommended, because it demonstrated a shorter incubation period (24 hours) and ready availability of the components required. Not all laboratory workers immediately accepted the suggested medium. However, at present, in the United States, only the New York City Public Health Laboratory uses a different basal medium. Their medium is similar to the modified McLeod's agar with additional enrichments and inhibitors.

With the establishment of a recommended culture medium came the massive use of penicillin and the dramatic effect it displayed against the gonococcus. Many authorities felt that gonorrhoea could be eradicated through early treatment schedules. A system known as speed zone epidemiology was introduced in 1952 in 10 cities in the United States. Male patients with gonorrhoea were interviewed to trace relevant contacts and every effort was made to treat contacts within 72 hours of their identification. In most clinics contacts received treatment during the prescribed period. However, the number of males with gonorrhoeal urethritis attending the clinics remained essentially the same. In 1956 an approach called antibiotic quarantine was combined with speed zone epidemiology. This was a combination of rapid treatment and use of benzathine penicillin to maintain prolonged blood levels, in an effort to prevent reinfection for a period of two to six weeks. Despite the enthusiastic application of this approach for several years, it became apparent that the incidence of gonorrhoea had not been significantly reduced.

During this period of the 1950s, the use of culture to aid in the identification of the gonococcus in most public health laboratories fell to almost zero. Some renewed interest was displayed with the introduction of the immunofluorescence methods for detection and identification of the gonococcus in urethral exudates [2]. The chief advantages of the fluorescent antibody procedures over the culture method were the time required to make a definite diagnosis – 16–20 hours as against two to ten days – and the greater sensitivity of the technique in most laboratories.

In 1964, Thayer and Martin described a combination of additives with selective properties used in a conventional culture medium which permitted growth of pathogenic Neisseria while greatly inhibiting or suppressing the saprophytic Neisseria species [3]. Overgrowth of gonococcal colonies by bacterial contaminants encountered in cervical, vaginal and anal canal specimens was almost totally prevented. This selective medium was prepared by adding 25 units/ml of polymyxin B and 10 μg/ml of ristocetin to the previously sterilized conventional culture medium.

Although field trials indicated many advantages of the selective medium, several serious deficiencies also appeared. Firstly, inhibition of some gonococci resulted in false-negative reports and, secondly, to overcome this inhibitory effect the incubation period had to be extended from 24 to 48 hours.

The withdrawal of ristocetin from the commercial market prompted a search for an alternative antibiotic to inhibit Gram-positive organisms. Vancomycin at a concentration of 3.0 μg/ml was selected after testing of 57 recent isolates revealed no inhibition at the 7.5 μg/ml level.

It was also noted that colistin mesilate sodium proved to be more effective against the saprophytic Neisseria than polymyxin B. The incorporation of nystatin, in addition, resulted in the inhibition of some yeast strains present in vaginal specimens. An additional effect of nystatin was its inhibitory effect against certain airborne moulds that were a constant problem in the medium preparation area. The new combination of vancomycin, colistin and nystatin (VCN) therefore showed promise [4].

Gonococcal growth on the new VCN combination became evident earlier and the saprophytic Neisseria and micrococci were completely inhibited. A medium with a high degree of specificity and selective sensitivity had therefore been attained, making it possible to establish a culture testing programme for gonorrhoea with more assurance than previously.

Plans were being developed during the late 1960s for a national programme of gonorrhoea control. The sensitivity of the VCN medium to be used in that programme was established by three groups of investigators. Lucas and co-workers [5], Schmale et al. [6] and Caldwell et al. [7] reported that cervical specimens taken for culture from a total of 1,354 patients resulted in sensitivities of 92.4 %, 93.8 % and 96 % respectively.

Although these studies showed the selective medium to be a usable diagnostic tool in the larger population centres with modern laboratory facilities, a tool suitable for small towns, cities and rural areas did not exist. The development of Transgrow therefore followed. This new system containing VCN medium plus trimethoprim, dispensed in a bottle, under a blanket of increased carbon dioxide became a useful tool for those who had a small workload and little or no access to laboratory facilities [8].

Hosty and co-workers [9] compared the effectiveness of holding specimens for 24 hours using five different transport media. The results with the unincubated Transgrow and the enriched cystine trypticase agar (CTA), held 24 hours before incubation, were good, while the other three transport media were inadequate for the preservation of gonococci. When the holding time was extended to 48 hours, the three media tested showed considerable losses. Of these, Transgrow was best.

The widespread acceptance and use of Transgrow brought with it many unexpected problems. A study was, therefore, initiated to correct the weak areas. The first step was to formulate a tablet which would release an amount of CO_2 sufficient to satisfy the requirements of growing gonococci. This was accomplished by combining sodium bicarbonate and citric acid in amounts that would generate 20.0 ml of carbon dioxide. An inoculated Petri dish was then placed in a plastic bag with the tablet and sealed [10]. The medium contains water which causes the humidity to rise and induces a reaction in the tablet. In two hours a 5 % carbon dioxide atmosphere, necessary for growth of the gonococcus, is generated. Efforts to improve this system continued as it did not give complete satisfaction.

The result of those efforts is the Biological Environmental Chamber (BEC) [11]. Some advantages the chamber offers are that the CO_2 generating source is contained within the culture system, that the system is more convenient for transportation and that it allows the laboratory worker easy access to the culture for further examination and testing.

An evaluation by Jephcott and colleagues [12] comparing the BEC method with their conventional methods in 167 paired specimens from female contacts detected four additional positives: there were 73 positives using the conventional tests while the BEC method detected 77. However, the BEC method failed to grow 11 cultures that were positive by the conventional methods. The conclusions of these workers were that the BEC method was accurate, reliable and less time-consuming than conventional methods and its use was suggested in laboratories where more definite but more complex tests are unavailable.

The increased use of the selective medium in the National Gonorrhoea Control

Programme has identified some disadvantages, predominant among which is the relative ineffectiveness of nystatin in inhibiting the growth of yeast contaminants. The inadequacy is due to the short life of nystatin in the culture medium. To have an effective culture programme it is usually necessary to prepare and store large quantities of medium. In addition, several investigators have shown that growth of the gonococcus is inhibited by the presence of *Candida albicans* [13, 14], resulting in overgrowth or masking of a positive culture. Another problem associated with the use of selective medium is the lack of inhibition of staphylococci by vancomycin at a concentration of 3.0 $\mu g/ml$. A recent report [15] describes the substitution of 20.0 $\mu g/ml$ anisomycin, a water-soluble antimycotic agent, for nystatin and the increase of the vancomycin concentration to 4.0 $\mu g/ml$. A dramatic inhibition of yeast growth and a reduction of approximately 75 % in staphylococci was noted. Two studies have suggested that these changes in the formulation do not affect the gonococcus [16, 17].

Selective culture media are seldom absolute. However, the results of these studies of the new combination of antimicrobials suggest almost total inhibition of concomitant micro-organisms with no detectable effect on the growth of the gonococcus. If field trials continue to support the findings in this recent work, it is the opinion of many investigators that we probably will have reached the upper limit in a single-culture detection system for the gonococcus.

Penicillinase-producing *Neisseria gonorrhoeae* (PPNG) strains have been isolated and confirmed in many parts of the world. Descriptions have been given of several rapid laboratory tests which are extremely useful in detecting β-lactamase production by cultures of *N. gonorrhoeae*. As presently designed, none of these tests can be performed directly on clinical specimens.

Recently, a primary selective culture medium for detecting PPNG has been reported. Biplates were prepared with one half of each containing the inhibitors vancomycin (4 $\mu g/ml$), colistin mesilate sodium (7.5 $\mu g/ml$), the new antimycotic agent, anisomycin, (20 $\mu g/ml$) and trimethoprim (5 $\mu g/ml$) (VCAT). The other half contained the same concentrations of the same inhibitors but penicillin G (2.5 units/ml) or ampicillin (2.5 $\mu g/ml$) was also added and 2 % heat-inactivated horse serum was substituted for the 1 % haemoglobin. In addition, 2.0 ml of a heavy suspension of *Sarcina lutea* (Food and Drug Administration strain number 1001) susceptible to < 0.01 $\mu g/ml$ of penicillin was added to 100 ml of the molten (45 °C) agar and mixed before the second half of the biplate was filled.

Using this medium 8.1 % more PPNG isolates were identified than with the established chemical tests for β-lactamase. If these results are supported by field trials now under way, this medium could be used to provide therapeutically useful information for the clinician within 18–24 hours and also to confirm the presence of β-lactamase-producing cultures for the microbiologist, clinician and epidemiologist within 72 hours from the time the clinical specimen is streaked on the selective culture medium.

SUMMARY

The role of laboratory media for the investigation of gonococci utilized in the United States over the past 40 years is reviewed. The evolution of advances in media for the identification and transport of gonococcal specimens is discussed. These advances have placed clinical and public health laboratories in the forefront of the vigorous campaign to control gonorrhoea.

REFERENCES

1. Carpenter, C.M., Bucca, M., Buck, T., Casman, E., Christensen, O., Drowe, E., Drew, R., Hill, J., Lankford, C., Morton, H., Peizer, L., Shaw, C. and Thayer, J. (1949): Evaluation of twelve media for the isolation of the gonococcus. *Am. J. Syphilitic, Gonococcal and Vener. Dis., 33,* 164.
2. Deacon, W., Peacock, W., Freeman, E. and Harris, A. (1959): Identification of *Neisseria gonorrhoeae* by means of fluorescent antibodies. *Proc. Soc. Exp. Biol. Med., 101,* 322.
3. Thayer, J. and Martin, J. (1964): A selective medium for the cultivation of *N. gonorrhoeae* and *N. meningitidis. Public Health Rep., 79,* 49.
4. Thayer, J. and Martin, J. (1966): Improved medium selective for the cultivation of *N. gonorrhoeae* and *N. meningitidis. Public Health Rep., 81,* 559.
5. Lucas, J., Price, E. and Thayer, J. (1967): Diagnosis and treatment of gonorrhoea in the female. *N. Engl. J. Med., 276,* 1454.
6. Schmale, J., Martin, J. and Domescik, G. (1969): Observations on the culture diagnosis of gonorrhea in women. *J. Am. Med. Assoc., 210,* 312.
7. Caldwell, J., Price, E., Pozin, G. and Cornelius, C. (1971): Sensitivity and reproducibility of Thayer-Martin culture medium in diagnosing gonorrhea in women. *Am. J. Obstet. Gynecol., 109,* 463.
8. Martin, J. and Lester, A. (1971): Transgrow, a medium for transport and growth of *N. gonorrhoeae* and *N. meningitidis. Health Services and Mental Health Administration Health Reports, 86,* 30.
9. Hosty, T., Freear, M., Baker, C. and Holston, J. (1974): Comparison of transportation media for culturing *N. gonorrhoeae. Am. J. Clin. Pathol., 62,* 435.
10. Martin, J., Armstrong, J. and Smith, P. (1974): New system for cultivation of *Neisseria gonorrhoeae. Appl. Microbiol., 4,* 802.
11. Martin, J. and Jackson, R. (1975): A biological environmental chamber for the culture of *Neisseria gonorrhoeae. J. Am. Vener. Dis. Assoc., 2,* 28.
12. Jephcott, A., Bhattacharyya, M. and Jackson, D. (1976): Improved transport and culture system for the rapid diagnosis of gonorrhoea. *Br. J. Vener. Dis., 52,* 250.
13. Hipp, S., Lawton, W., Chen, N. and Gaafar, H. (1974): Inhibition of *N. gonorrhoeae* by a factor produced by *C. albicans. Appl. Microbiol., 27,* 192.
14. Hipp, S., Lawton, W., Savage, M. and Gaafar, H. (1975): Selective interaction of *N. gonorrhoeae* and *C. albicans* and the possible role in clinical specimens. *J. Clin. Microbiol., 1,* 476.
15. Martin, J. and Lewis, J. (1977): Anisomycin: improved antimycotic activity in modified Thayer-Martin medium. *Public Health Lab., 35,* 53.
16. Moyer, N. and Parsons, B. (1977): Clinical evaluation of a new medium for isolation of gonococci. *Public Health Lab., 35,* 196.
17. Smeltzer, M., Curran, J. and Lossick, J. (1979): A comparative evaluation of media used to culture *N. gonorrhoeae. Public Health Lab., 37,* 43.
18. Martin, J. and Lewis, J. (1977): Screening *Neisseria* for penicillinase production. *Lancet, I,* 1014.

Discussion

J. A. Morello (Chicago, U.S.A.): Were the 8.1 % more PPNG that were isolated on this medium then tested chemically to determine whether they produced β-lactamase?

J. S. Lewis: This has happened in many situations in which PPNGs have been isolated and identified chemically in any number of different places and then sent to CDC for confirmation only to be found negative in subsequent testing. The laboratories where the PPNGs are found are good laboratories. Perhaps Dr Morello has had some?

J. A. Morello: We have had some real ones!

J. S. Lewis: The plasmid is very unstable and few in number in some isolates. Like Dr Lind, again, in some isolates we can lyophilize and retain the plasmid but in others we lyophilize and the plasmid has gone.

E. Sandström (Stockholm, Sweden): Would Dr Lewis expect all field isolates to grow on the clear medium?

J. S. Lewis: We have no indication, nor reason to expect that they would not. The biggest difference – in fact, *the* difference – is the substitution of horse plasma for haemoglobin.

A review of techniques for the identification of *Neisseria gonorrhoeae* and of new developments

S. A. Waitkins

Public Health Laboratory, Fazakerley Hospital, Liverpool, England

INTRODUCTION

The isolation of *Neisseria gonorrhoeae* by cultural methods is essential if a correct medicolegal diagnosis of the disease is to be made. The findings of typical intracellular, Gram-negative diplococci among pus cells from genital discharge may alert the clinician to start antibiotic treatment but full identification rests on the ability of the microbiology laboratory to isolate the organism and perform confirmatory tests.

In recent years, other organisms morphologically similar to the gonococcus have frequently been isolated from the genital tract, pharynx and rectum. These organisms include both *Neisseria meningitidis* and *Neisseria lactamica*. It is, therefore, most important that organisms are successfully isolated and further investigations carried out to confirm the original clinical diagnosis. It is also important to establish whether the patient has been cured. It is our practice in the United Kingdom to record three negative culture findings over a period of three weeks before accepting that treatment has been effective. Finally, isolation of organisms enables the laboratory to monitor the antibiotic sensitivity patterns of the gonococcus. It was by such vigilance that the penicillinase-producing gonococci were first noticed [1–3].

The argument for isolating and subsequently identifying *N. gonorrhoeae* is now well established and recognized in most countries.

COLLECTION

The most important stage in the procedure for isolation of the gonococcus, if this is to be successful, is the collection of the specimen by the clinician. No matter how sophisticated the available laboratory techniques, without accurate sampling many false-negative cultures result. It must be emphasized that the correct clinical specimen from males is the *urethral swab*, not a swab from the meatus, and that in the homosexual pharyngeal as well as rectal samples should be taken. In females the cervix is the most reliable site (Table I). However, Wilkinson [4], in a study covering 368 patients, showed that in 19 % of these patients only one of the four sites from which samples were taken was positive.

Table I. Frequency of isolation from four sites in 368 women with gonorrhoea. Reprinted with permission from Wilkinson [4].

Site	Percent of women	
	giving positive cultures	giving positive cultures only at site indicated
Urethra	70	3
Cervix	87	9
Vagina	71	4
Rectum	36	3

A single set of negative cultures need not necessarily exclude the diagnosis, as shown by the data of Thin and Shaw [5] (Table II).

Extragenital infections with *N. gonorrhoeae* have become much more prevalent in the last 10 years. Orogenital sexual practices may result in acute pharyngitis, primarily among women and male homosexuals. Extragenital infections are often asymptomatic and can be extremely difficult to detect [6, 7]. However, if a suction device similar to that described by From and Veien [8] is used, improved isolation rates are achieved. Post-gonococcal septicaemia is usually uncommon but can occur [9], while skin lesions are rarely positive but organisms can be demonstrated by fluorescent antibody techniques [4]. On the other hand, indirect extragenital gonococcal infection can be seen in newborn infants suffering from ophthalmia neonatorum contracted during their passage through an infected birth canal. In this condition it is most important that the causative agent is correctly bacteriologically identified.

TRANSPORTATION

Regardless of the number of swabs taken, or how well they are collected, their transportation to the laboratory may present many difficulties.

Table II. Cases diagnosed at first, second and third examinations. Reprinted with permission from Thin and Shaw [5].

Examination	Positive results by			
	Gram stain or culture or both		Gram stain	
	Number	%	Number	%
First	138	95.2	75	54.3 (of 138)
Second	5	3.4	3	60 (of 5)
Third	2	1.4	2	100 (of 2)
Total	145	100.0		

Direct plating on selective and/or non-selective media followed by immediate incubation in a humid atmosphere containing 5 % CO_2 is ideal. However, if this is not possible, and delay in transportation of the clinical specimen to the laboratory is expected, a holding medium must be used. A medium in common use is Stuart's transport medium [10]. Modifications such as carbon particle incorporation [11] and increased agar concentration [12] have improved isolation rates. Recent developments maintaining the medium's holding capacity but, at the same time, allowing limited growth of the gonococcus have proved fairly successful. Media like Transgrow [13] which allows growth on selective medium in a pregassed bottle containing 10 % CO_2 can be very easily inoculated in the venereal disease (V.D.) clinic and then posted to the laboratory. A problem with Transgrow is that condensation of excess water on the side of the bottle occurs rapidly, seriously affecting the visibility of individual colonies. Both JEMBEC/Neigon® and Microcult-GC-Test® overcome these problems by biochemical addition of CO_2 and facilitation of access to the organisms [14, 15]. We have found the GC-Till-U-Test® to be very satisfactory, particularly if long postal delays are anticipated [16]. This test makes use of a double-sided plastic container containing Thayer-Martin selective medium. The clinician can plate directly on the medium and post the container to the laboratory. CO_2 is generated by a bicarbonate tablet. Using the GC-Till-U-Test® we have demonstrated that clinical isolates can survive for up to 72 hours before isolation rates are seriously affected. This would certainly prove to be most useful in clinics distant from main regional laboratories.

ISOLATION

Once the clinical specimen containing the gonococci reaches the laboratory, isolation procedures may be applied. The difficulty of isolating gonococci from specimens is that they are invariably contaminated by the normal flora of the sites sampled and these contaminants quickly outgrow the more fastidious gonococci if no attempt at active selection is made. To this end, many laboratories use selective media incorporating the original cocktail of antibiotic described by Thayer and Martin [17]: vancomycin 3 $\mu g/ml$, colistin 12.5 $\mu g/ml$ and nystatin 12.5 units/ml. Proteus species may present a particular problem in rectal swabs. These may be suppressed by the addition of trimethoprim at a concentration of 5 $\mu g/ml$ [18]. There have been suggestions [19] that vancomycin at the concentration used suppresses the growth of about 4 % of gonococci. Lincomycin (0.5 $\mu g/ml$) may be used as an alternative [20] but, in general, we find that overgrowth of contaminants is increased using lincomycin and, consequently, still prefer vancomycin.

Many basic growth media have been used in conjunction with the above mixture of antibiotics. The most common are that of Martin and his colleagues, known as Iso-vitalex® (BBL) [21] and the original clear medium of Kellogg (Difco) [22]. Both give very satisfactory growth and are readily available commercially.

Ideally, of course, both selective and non-selective media should be inoculated simultaneously, thus ensuring that most of the potential gonococcal isolates are recovered. However, in a busy routine laboratory this may not be done and great reliance must then be placed on experienced technical interpretation of the selective plates.

CONFIRMATORY TESTS

There are three main types of confirmatory test that can be performed by the laboratory:
1. sugar fermentation,
2. co-agglutination,
3. fluorescent antibody.

In the United Kingdom there still exists a statutory demand for *N. gonorrhoeae* to be confirmed. The only test that satisfies requirements is sugar fermentation.

Sugar fermentation tests can be categorized in types:
a) routine agar base + 10 % animal tissue fluid supplement media,
b) serum-free media,
c) rapid fermentation.

Routine methods using 10 % rabbit serum or hydrocele fluid added to either proteose peptone bases or nutrient agar plus phenol red as pH indicator give satisfactory results. However, it is generally believed that rabbit serum may contain minute amounts of maltase, thereby giving a false-positive reading in the maltose reaction. Alternatively, small amounts of the maltose sugar may be present in the glucose as a contaminant giving a false-positive result with glucose fermentation. The availability of hydrocele fluid for use in sugar fermentations is extremely limited. It cannot easily be obtained and therefore is used only by a few very specialized laboratories. To overcome the disadvantages with tissue fluid-based media, serum-free media were developed [23, 24]. The serum-free media contain nutrient broth agar with cysteine, ferric nitrate and cocarboxylase supplements to increase the growth of the organism. Both types of medium give satisfactory results which can easily be interpreted. Their only disadvantage is that they rely on the growth of the organism to metabolize the sugars. This may take up to 48 hours, which delays the reporting of results.

An alternative approach is to determine whether preformed enzymes exist. Kellogg and Turner [25] used this approach with a buffered salt solution. When a heavy inoculum of gonococci is added the resulting coloration of the sugar-containing solution indicates the positive reaction. Such methods are usually simple, inexpensive and quick, normally taking less than one hour [26, 27].

Co-agglutination consists of using *Staphylococcus aureus* with antibodies to the gonococcus absorbed on its protein A and allows differentiation of gonococci from other Neisseria [28]. Variations in this test have been recorded by many observers. These may be due to local antigenic variation in *N. gonorrhoeae*. A co-agglutination method has been recently marketed by Pharmacia Diagnostics AB. As yet there has been little assessment of this kit but preliminary results seem encouraging. Menck [29] found that cross-reactions between *N. meningitidis* and *N. gonorrhoeae* can exist but may be eliminated by growing the test organism on serum-free media.

The fluorescent antibody (FA) test is by far the most popular rapid confirmation test available today. The FA conjugate must be highly specific for *N. gonorrhoeae* and must not show fluorescence with any of the other Neisseria. To achieve this degree of specificity, tedious absorption methods must be used to obtain conjugate. Even after such lengthy procedures many commercially available FA conjugates show a considerable amount of fluorescence with both *N. meningitidis* and *N.*

Table III. Penicillinase production by Neisseria gonorrhoeae and variation in fluorescent antibody reaction.

Organism	Penicillin M.I.C. (Units/ml)	Commercially prepared conjugate	Laboratory prepared conjugate
Non-penicillinase-producing *Neisseria gonorrhoeae*: Freshly isolated			
78/68350	0.5	+ +	+ + +
78/68352	0.06	+ + +	+ + +
78/68660	0.125	+ + +	+ + +
78/69490	0.03	+ + +	+ + +
78/70463	0.008	+ +	+ + +
Preserved in liquid nitrogen			
78/4358	0.05	+ +	+ + +
78/15291	0.06	+ +	+ + +
78/15713	0.015	+ +	+ + +
78/15715	0.125	+ + +	+ + +
78/16149	0.008	+ + +	+ + +
78/15151	0.008	+ +	+ + +
Penicillinase-producing *Neisseria gonorrhoeae*: Freshly isolated			
78/5414	> 100	+	+ +
78/7688	> 100	−	+ +
78/72767	> 100	±	±
78/74163	> 100	+ +	+ + +
Preserved in liquid nitrogen			
76/16353	> 100	−	+ +
76/25777	> 100	+ +	+ + +
76/38401	> 100	±	±
76/78980	> 100	+ + +	+ + +
AP 672	> 100	+	+ + ±
Negative control (*N. meningitidis*)		−	−

lactamica preparations. If only the FA test is used there is a danger of misdiagnosis. The FA test must always be supplemented by biochemical testing if definite proof of gonococcal infection is requested and in any case in extragenital infections.

FA methods are subject not only to interference from commensal Neisseria but, in some rare instances, may even miss gonococcal organisms. We have found that commercial preparations occasionally give false-negative results with penicillinase-producing gonococci. Table III lists our findings. In some instances the fluorescence reaction was severely reduced while in two cases (78/7688 and 76/16353) negative results were recorded. Neither the antigen used to raise antibody in the case of commercial conjugate nor that used to prepare conjugate in the laboratory contained penicillinase-producing gonococci. However, in the laboratory-prepared conjugate, the gonococci used were more recently isolated, this being reflected in their increased fluorescence reaction. This property of penicillinase-producing gonococci is intriguing. Further studies are required to investigate the antigenic structure of these organisms [30].

SUMMARY

The ability to isolate and identify *N. gonorrhoeae* reliably is most important not only from the point of view of good bacteriological practice but also for epidemiological and, in some countries, medicolegal reasons. The isolation procedures are fairly well established and can be accomplished using the various commercially available selective media. Identification methods, on the other hand, are currently undergoing re-appraisal. Classical sugar fermentation methods are no longer feasible in busy routine laboratories and more rapid tests need to be developed. Such rapid tests can employ either fermentation methods or detect the organism by specific antibodies. Both co-agglutination and fluorescent antibody tests work satisfactorily but the danger of misdiagnosis when these rapid methods are used must always be remembered.

REFERENCES

1. Phillips, I. (1976): β-lactamase producing, penicillin resistant gonococcus. *Lancet, II,* 656.
2. Percival, A., Rowlands, J., Corkhill, J.E., Alergant, C.D., Arya, O.P., Rees, E. and Annels, E.H. (1976): Penicillinase-producing gonococci in Liverpool. *Lancet, II,* 1379.
3. Turner, G.C., Ratcliffe, J.G. and Anderson, R.D. (1976): Penicillinase-producing *Neisseria gonorrhoeae. Lancet, II,* 793.
4. Wilkinson, A.E. (1977): Cultural methods for the diagnosis of gonorrhoea. In: *Gonorrhoea: Epidemiology and Pathogenesis,* FEMS symposium No. 2. Editors: F.A. Skinner, P.D. Walker and H. Smith. Federation of European Microbiological Societies, Amsterdam.
5. Thin, R.N. and Shaw, E.J. (1979): Diagnosis of gonorrhoea in women. *Br. J. Vener. Dis.,* 55, 10.
6. Fowler, W. (1974): Gonorrhoea of the pharynx. *Br. Med. J.,* 2, 440.
7. Bro-Jørgenson, A. and Jenson, T. (1973): Gonococcal pharyngeal infections. Report of 110 cases. *Br. J. Vener. Dis.,* 49, 491.
8. From, E. and Veien, N.K. (1974): Tonsillar gonorrhoea demonstrated by a suction device. *Br. J. Vener. Dis.,* 50, 360.
9. Gelpi, A.P. (1974): Gonococcal sepsis in college students. *J. Am. Coll. Health Assoc.,* 23, 157.
10. Stuart, R.D., Toshack, S.R. and Patsula, T.M. (1954): Problem of transport of specimens for culture of gonococci. *Can. J. Publ. Health,* 45, 73.
11. Amies, C.R. (1967): A modified formula for the preparation of Stuart's transport medium. *Can. J. Publ. Health,* 58, 296.
12. Gastrin, B. and Kallings, L.O. (1968): Improved methods for gonococcal sampling and examination on a large scale. *Acta Pathol. Microbiol. Scand.,* 74, 362.
13. Martin, J.E. and Lester, B.A. (1971): Transgrow, a medium for transport and growth of *Neisseria gonorrhoeae* and *Neisseria meningitidis. HSMHA Health Reports,* 86, 30.
14. Jephcott, A.E., Bhattacharyya, M.N. and Jackson, D.H. (1976): Improved transport and cultural system for the rapid diagnosis of gonorrhoea. *Br. J. Vener. Dis.,* 52, 250.
15. Unsworth, P.F., Talsania, H. and Phillips, I. (1979): An assessment of the Microcult-GC culture test. *Br. J. Vener. Dis.,* 55, 1.
16. Waitkins, S.A. and Anderson, R.D. (1979): Personal communication.
17. Thayer, J.D. and Martin, J.E. (1964): A selective medium for the cultivation of *Neisseria gonorrhoeae* and *Neisseria meningitidis. Public Health Rep.,* 79, 49.

18. Seth, A.D. (1970): Use of trimethoprim to prevent overgrowth by *Proteus* in the cultivation of *N. gonorrhoeae*. *Br. J. Vener. Dis.*, *46*, 201.
19. Reyn, A. and Bentzon, M.W. (1972): Comparison of a selective and a non-selective medium in the diagnosis of gonorrhoea to ascertain the sensitivity of *Neisseria gonorrhoeae* to vancomycin. *Br. J. Vener. Dis.*, *48*, 363.
20. Odegaard, K., Solberg, O., Lind, J., Myrhe, G. and Myland, B. (1975): Lincomycin in selective medium for the isolation of *Neisseria gonorrhoeae*. *Acta Pathol. Microbiol. Scand. Sect. B*, *83*, 301.
21. Martin, J.E., Billings, T.E., Hackney, J.F. and Thayer, J.D. (1967): Primary isolation of *Neisseria gonorrhoeae* with a new commercial medium. *Public Health Rep.*, *82*, 361.
22. Kellogg, D.S., Peacock, W.A., Deacon, W.E., Brown, L. and Pirkle, C.I. (1963): *Neisseria gonorrhoeae*. 1. Virulence genetically linked to clonal variation. *J. Bacteriol.*, *85*, 1274.
23. Flynn, J. and Waitkins, S.A. (1972): A serum-free medium for testing fermentation reactions in *Neisseria gonorrhoeae*. *J. Clin. Pathol.*, *25*, 525.
24. Wallace, R., Ashton, F., Charron, F. and Diena, B.B. (1975): An improved sugar fermentation technique for the confirmation of *Neisseria gonorrhoeae*. *Can. J. Publ. Health*, *66*, 251.
25. Kellogg, D.S. and Turner, E.M. (1973): Rapid fermentation confirmation of *Neisseria gonorrhoeae*. *Appl. Microbiol.*, *25*, 550.
26. Brown, W.J. (1974): Modification of the rapid fermentation test for *Neisseria gonorrhoeae*. *Appl. Microbiol.*, *27*, 1027.
27. Morse, S.A. and Bartenstein, L. (1976): Adaptation of the Minitek system for rapid identification of *Neisseria gonorrhoeae*. *J. Clin. Microbiol.*, *3*, 8.
28. Danielsson, D. and Kronvall, G. (1974): Slide agglutination method for the serological identification of *Neisseria gonorrhoeae* with anti-gonococcal antibodies absorbed to protein A containing staphylococci. *Appl. Microbiol.*, *27*, 368.
29. Menck, H. (1976): Identification of *Neisseria gonorrhoeae* in cultures from tonsillo-pharyngeal specimens by means of a slide co-agglutination test (Phadebact Gonococcus Test). *Acta Pathol. Microbiol. Scand. Sect. B*, *84*, 139.
30. Waitkins, S.A. and Anderson, R.D. (1979): The variation of the fluorescent antibody reaction between penicillinase- and non-penicillinase-producing gonococci. Presented at the Pathological Society of Great Britain and Ireland, London, January 1979.

Discussion

I. Lind (Copenhagen, Denmark): What is the percentage of overgrown cultures in Dr Waitkins' laboratory using either vancomycin or lincomycin?

S. A. Waitkins: I cannot give absolute figures offhand but using vancomycin about 12–13 % of the plates become overgrown, although they can be read because we have experienced technicians. With lincomycin it is much more, about 25 %.

I. Lind: During the last two months, when we have added lincomycin to our selective medium, the percentage has been 0.3 to 0.4.

S. A. Waitkins: It must be remembered that we are a routine laboratory, doing 70–100 specimens a day.

I. Lind: We are a routine laboratory, too, but we are handling 900–1200 specimens a day. One explanation for the difference may be that the standard of the media used in Denmark is very good.

S. A. Waitkins: Perhaps that is the reason.

I. Lind: I would like to comment on the immunofluorescence test. If it is done as a routine, new antisera must be prepared about every second year using freshly isolated strains for the immunization. This is to ensure that the reagent used for the identification contains antibodies analogous to the antigens of the gonococcal strains circulating at that time. Doing so, we have had no problems with the identification of either penicillinase-producing strains or strains with high chromosomal resistance. I suspect that the conjugate you used in this study was that prepared by you in our laboratory in 1974.

S. A. Waitkins: That is also true, but it is an indication perhaps of how bad the commercial preparations are. People in routine laboratories do not have the chance of preparing their own conjugates. We are at the mercy of the commercially-prepared materials.

I. Lind: Yes.

Comparison of the cystine-tryptic digest agar-carbohydrate co-agglutination and BACTEC® Neisseria differentiation methods for identification of *Neisseria gonorrhoeae* in the clinical laboratory

J. A. Morello, S. Beheshti and M. Bohnhoff
Clinical Microbiology Laboratories, The University of Chicago Hospitals, Chicago, U.S.A.

Gonorrhoea has become the most common communicable bacterial infection in the United States, occurring in an estimated 3 million individuals each year [1]. This disease is widespread in other countries as well. As a result, there has been increased interest in developing accurate and rapid methods for isolating and identifying *Neisseria gonorrhoeae* and for distinguishing this organism from saprophytic Neisseria and *Neisseria meningitidis* which may, at times, share the same ecologic niches. Identification methods currently available include assays for carbohydrate degradation that depend on growth [2], pre-formed enzyme [3] or radiometric detection [4]. Identification can also be made by serologic methods such as fluorescent antibody [5, 6] and co-agglutination by gonococcal antibody bound to *Staphylococcus aureus* protein A [7, 8].

In the United States, the standard method for identifying Neisseria is to determine their ability to produce acid from carbohydrates in a cystine-tryptic digest agar (CTA)-base medium. However, laboratory workers often have difficulty obtaining results with this medium and reports may be delayed for three or more days. For those laboratories that screen blood cultures with the BACTEC® radiometric system [9], a Neisseria differentiation kit is available that provides results only three hours after the test vials are inoculated. The co-agglutination method, in theory, could provide even more rapid results if the test could be performed directly with organisms growing on the culture plate, since agglutination is visible within one or two minutes after the reagent is mixed with the organisms. We evaluated CTA-carbohydrate, BACTEC® and co-agglutination systems to determine their accuracy for identifying *N. gonorrhoeae* strains and for distinguishing them from other Neisseria species.

MATERIALS AND METHODS

The three methods were used to examine 180 strains of Neisseria species. Seventy-five were stock strains and 105 were fresh clinical isolates received either from patients at the University of Chicago Hospitals and Clinics or from the Howard Brown Memorial Clinic, a clinic on Chicago's north side frequented primarily by male homosexuals. Stock cultures were stored at $-80\,°C$ after only one or two transfers of the primary isolate. They were frozen in a 50% solution of Trypticase soy broth (BBL) and heat-inactivated horse serum for periods of from one month to up to nine years.

Stock cultures were transferred at least twice onto chocolate agar before they were examined and were incubated each time for 18 hours at 35 °C in a $7\% CO_2$ atmosphere. After the final incubation period, the organisms were tested by the three procedures. The CTA medium consisted of CTA semi-solid agar base (BBL) containing a final concentration of 1% filter-sterilized glucose, maltose and fructose and a control tube with no carbohydrate. This medium was prepared in our laboratory and pipetted into 13×100 mm screw-capped test tubes to a depth of approximately 3 cm. This contrasts with commercially-prepared CTA medium in which there is usually a high column of semi-solid fluid in a larger tube. To inoculate the tubes, a full 3 mm loopful of growth was scraped from the plate and inserted approximately 5 mm below the surface of each agar deep. The o-nitrophenyl-β-D-galactopyranoside (ONPG) test was substituted for a lactose fermentation tube to detect β-galactosidase. All tubes were examined after one, two, three and 24 hours' incubation at 35 °C in the absence of CO_2.

To prepare inoculum for the BACTEC® Neisseria Differentiation kit (Johnston Laboratories), a heavy suspension of organisms was made in the broth provided. A 0.3 ml aliquot of the suspension was inoculated into each of three vials containing ^{14}C-glucose, ^{14}C-maltose or ^{14}C-fructose. After three hours of incubation at 35 °C, the vials were monitored on a BACTEC® 460 instrument (Johnston Laboratories) to detect released $^{14}CO_2$ from the carbohydrates. For this test, a detectable value of 20 or above on the BACTEC® is considered to be a positive result. With our instrument, glucose readings commonly were 999, the upper limit of values, and maltose readings were approximately 200–300.

The co-agglutination test was performed with the Phadebact® Gonococcus Test (Pharmacia Diagnostics AB). Agglutination was scored as 0 to 3 + with 0 being clearly negative, and 1 + being a slight, fine agglutination that is considered to be a non-interpretable reaction. Agglutination of 2 + and 3 + was considered positive only when the control test was either 0 or 1 +, respectively. At the beginning of this study, the co-agglutination test was performed directly from colonies growing on the chocolate agar plate. Whenever possible, colonies conforming to Kellogg's description of colony types 1 and 2 [10] were selected, since these predominate in fresh, clinical cultures. A majority of the reactions, however, were non-interpretable, because there was as good agglutination with the control reagent as the test reagent.

In accordance with the manufacturer's directions, suspensions were then prepared in 0.5 ml of water from several colonies and boiled for 20 minutes before the co-agglutination procedure was performed. During the latter part of the study, the volume of water was decreased to 0.2 ml and the suspensions boiled for only five minutes. In a few instances, with N. gonorrhoeae isolates, strong agglutination persisted with the control reagent, but boiling the suspensions for an additional five minutes resolved the problem. All data presented here were obtained with boiled suspensions of the organisms.

The procedure for testing clinical isolates was similar to that for stock cultures. However, if the primary culture plate (usually modified Thayer-Martin (MTM) or chocolate agar) had sufficient growth of suspected gonococci, a suspension was prepared for boiling and co-agglutination. Otherwise, a subculture was made to chocolate agar and the three systems were used for organism identification. Even if a co-agglutination test was performed from the primary culture plate, it was repeated the following day with growth from the subculture.

RESULTS AND DISCUSSION

Table I illustrates the characteristics of the stock strains of *N. gonorrhoeae*. The 50 strains, isolated from approximately equal numbers of females and males, were recovered from systemic, urogenital, rectal and pharyngeal sites and varied in their susceptibilities to penicillin from highly susceptible (≤ 0.015 μg/ml) to one penicillinase-producing *N. gonorrhoeae* (PPNG) that had a penicillin minimum inhibitory concentration (MIC) of 16 μg/ml. The auxotypes of these strains represent those seen most commonly in the Chicago area [11]. Eleven strains requiring arginine, hypoxanthine and uracil (AHU) were included, since strains of this auxotype grow slowly and are more often difficult to characterize biochemically.

Table I. Characteristics of 50 stock strains of Neisseria gonorrhoeae.

Sex of patients:	28 males
	22 females
Source of isolates:	15 systemic (12 blood, 3 joint)
	21 urogenital (11 cervix, 10 urethra)
	9 rectum
	5 pharynx
Penicillin susceptibilities (μg/ml):	19: ≤ 0.015
	10: $= 0.018$–0.06
	21: ≥ 0.06 (includes 1 PPNG)
Auxotypes:	16 zero (requiring none of 11 nutrients tested)
	15 proline-requiring
	11 arginine-hypoxanthine-uracil (AHU)-requiring
	6 arginine-requiring
	2 proline- and arginine-requiring

Seventy-eight clinical isolates of *N. gonorrhoeae* were tested. Approximately one-third of these strains were from females. The sources were primarily genital or anal, but there were two blood isolates and one vaginal isolate from a three-year-old child.

Table II illustrates the identification results for the stock and clinical isolates of *N. gonorrhoeae*. The CTA carbohydrate method identified 94 and 100% of the

Table II. Identification of Neisseria gonorrhoeae by three systems.

Type of culture	Number of strains	Number (per cent) identified by each system		
		CTA	BACTEC®	Co-agglutination
Stock	50	47 (94)	50 (100)	45 (90)[a]
Clinical isolate	78	78 (100)	78 (100)	77 (99)[a]
Total	128	125 (98)	128 (100)	122 (95)

[a] All negative reactions recorded as 1+

strains, respectively. Two of the three negative stock strains belonged to the slow-growing AHU auxotype. Two subcultures may not, therefore, have been sufficient to restore physiologic competence to these strains. In contrast, the third negative isolate belonged to the zero auxotype which grows quite rapidly and in this case, therefore, the inoculum may have consisted of many non-viable cells. The BACTEC® system identified 100% of both stock and clinical isolates. With the co-agglutination method, five strains were scored as negative because they agglutinated only weakly (i.e., had a 1+ reaction). Among these five strains, there was no predominant auxotype, site of isolation or penicillin susceptibility. Four of the strains had been frozen for more than three years but the other was a recent isolate. After this study was completed, the co-agglutination test was repeated with these five strains. Since increased experience with the co-agglutination system had been gained, these tests were rescored as positive but the reactions were very weakly positive. The results overall were 98% identified by CTA, 100% by BACTEC® and 95% by co-agglutination.

In Table II, all of the co-agglutination data for clinical isolates were obtained with organisms that had been subcultured at least once from the primary isolation plate. Table III presents data for 34 clinical isolates of *N. gonorrhoeae* that were tested twice, first on the primary selective culture plate, i.e., the initial MTM plate, and then, the following day, after subculture to chocolate agar. More positive co-agglutination reactions were obtained after subculture to chocolate agar than from the initial·culture plate. In the latter instance, a third of the strains reacted only

Table III. Co-agglutination of N. gonorrhoeae from primary selective culture and after subculture.

	Number of strains (%)	
	Primary culture on MTM agar[a]	After subculture on chocolate agar[a]
Positive reaction	22 (65)	33 (97)
Non-interpretable:		
reaction in control	9 (26)	
1+ only reaction in test	3 (9)	1 (3)
Total	34 (100)	34 (100)

[a] All reactions performed with boiled, aqueous gonococcal suspension

Table IV. Identification of Neisseria gonorrhoeae by CTA-carbohydrate: time for positive reaction.

Type of culture	Number of strains	Accumulative percentages				Per cent negative
		1 hour	2 hours	3 hours	24 hours	
Stock	50	44	78	88	94	6
Clinical isolate	78	75	92	97	100	0
Total	128	63	87	94	98	2

weakly with the test reagent. The reason for this difference is not yet clear but in a very small pilot study performed with gonococci subcultured to both MTM and chocolate agar, co-agglutination reactions were much weaker and more often non-interpretable on the antibiotic-containing medium. Since vancomycin and colistin are included in MTM and affect the cell wall of bacteria, they may alter the antigenic structure of some gonococci, even though they may not inhibit their growth.

Table IV illustrates in detail the percentage of positive CTA-glucose reactions obtained with *N. gonorrhoeae* at each time period. Although many laboratory workers consider the CTA-carbohydrate method to be very slow, by using a heavy inoculum and a small volume of culture medium positive results for more than half of the strains were seen after only one hour and 94% were positive after three hours, which compares favourably with the three hour BACTEC® results. These rapid reactions are most likely due to pre-formed enzymes in the organisms rather than to bacterial growth.

Although isolates belonging to the AHU auxotype are often difficult to identify biochemically, as indicated in Table V, nine of 11 (82%) stock AHU strains produced acid from glucose after two hours. Those that were 'glucose-negative' after this time remained negative after 24 hours. The clinical isolates examined in this study have not yet been auxotyped. However, about 10% of them have the small colony morphology characteristic of the AHU strain [11] and all of these produced acid from glucose by three hours.

To test the specificity of the three identification systems, 25 meningococcus strains were examined. These included 13 stock strains which were isolated predominantly from systemic or extrapharyngeal sites and 12 clinical strains that were primarily pharyngeal isolates. The stock isolates belonged to five meningococcal serogroups (B, C, Y, 29E and W-135). The results for identification of meningococci by the

Table V. Time for positive reaction of 11 stock AHU strains in CTA-carbohydrate medium.

	Positive in	Number	Per cent	Accumulative % positive
	One hour	3	27	27
	Two hours	6	55	82
	Negative	2	18	82
	Total	11	100	

Table VI. Identification of Neisseria meningitidis by three systems.

Type of culture	Number of strains	Number (per cent) identified by each system		
		CTA	BACTEC®	Cross-reaction with co-agglutination
Stock	13	12 (92)[a]	13 (100)	0
Clinical isolate	12	12 (100)	12 (100)	0[b]
Total	25	24 (96)	25 (100)	0

[a] Patient had received cefalotin and gentamicin; maltose only +, BACTEC® low, but positive glucose reading (61)
[b] Four strains non-interpretable; two auto-agglutinable; two = 1+ reaction

three systems are presented in Table VI. One of the stock strains (a group B meningococcus), isolated from a patient receiving cefalotin and gentamicin therapy, did not produce acid in the CTA-glucose medium. Although the BACTEC® glucose reading for this strain was positive, it was extremely low, at only 61, in contrast to a reading of 900 or more for most other strains. Otherwise, all of the strains were identified by the CTA and the BACTEC® methods. With the co-agglutination test there were no reactions that would have been scored as positive, although four of the clinical isolates gave non-interpretable results. Two of these strains were autoagglutinable and two gave a 1+ reaction.

Table VII illustrates the percentage of positive CTA-glucose and maltose reactions for *N. meningitidis* at each time period. The maltose reactions, especially with the clinical isolates, developed quite rapidly, usually by two hours, and most of the glucose reactions were also positive by this time. Reactions in the stock cultures were somewhat delayed, but at least two-thirds were complete by three hours and all that became positive were positive after 24 hours.

Table VIII illustrates results obtained with a variety of Neisseria species and two strains of *Branhamella catarrhalis*. These strains were primarily clinical rather than stock isolates. All strains of *Neisseria lactamica*, *Neisseria sicca* and *B. catarrhalis*

Table VII. Identification of Neisseria meningitidis by CTA-carbohydrate: time for positive reaction.

Type of culture	Number of strains	Accumulative percentages in time shown (hours)									
		Glucose					Maltose				
		1	2	3	24	% negative	1	2	3	24	% negative
Stock	13	23	54	69	92	8	39	70	85	100	0
Clinical isolate	12	58	91	91	100	0	75	100	100	100	0
Total	25	40	72	80	96	4	56	84	92	100	0

Table VIII. Identification of Neisseria species and B. catarrhalis by three systems.

Organism	Number of strains	Number (per cent) identified by each system		
		CTA + ONPG	BACTEC® + ONPG	Cross-reaction with co-agglutination
N. lactamica	10	10 (100)	10 (100)	5 (50)[a]
N. sicca	9	9 (100)	9 (100)	0[b]
B. catarrhalis	2	2 (100)	2 (100)	0
'Other' Neisseria[c]	6	All negative	Glucose-positive	All negative

[a] Two additional strains non-interpretable: one auto-agglutinable, one = 1 +
[b] All strains auto-agglutinable
[c] Other reactions: no growth on MTM; growth on nutrient agar; N_2 gas from NO_2

were identified by both the CTA and BACTEC® methods. *N. lactamica* was differentiated from *N. meningitidis* by the positive ONPG reaction of the former. In the co-agglutination test, however, half of the 10 *N. lactamica* isolates gave positive reactions, i.e., they agglutinated with the test reagent but not with the control reagent. If only co-agglutination had been performed, these isolates would have been identified as *N. gonorrhoeae*. All of the *N. sicca* strains were scored as non-interpretable, since they were auto-agglutinable. The two *B. catarrhalis* strains were clearly negative. No agglutination was observed. The last category, six 'other' Neisseria, may be *Neisseria flavescens*. Three of these were isolated from respiratory specimens, one was isolated from a cervix and two are probably colony variants of one strain isolated from the blood of an infant with signs and symptoms of meningococcaemia. All of these strains were isolated on blood or chocolate agar and were identified initially by the laboratory as *N. gonorrhoeae* on the basis of the positive glucose readings on the BACTEC®. The CTA-carbohydrate and co-agglutination reactions, however, were consistently negative. These isolates are not gonococci because they do not grow on MTM medium, do grow on nutrient agar and also produce N_2 gas from NO_2. Also, all of them retain their coccal morphology when exposed to sub-inhibitory concentrations of penicillin [12]. The coccobacillary Eikenella and Kingella can, therefore, also be excluded. Examination of the BACTEC® glucose readings for these strains indicated that all were quite low, ranging from 280–550, with the exception of one strain that had a reading of 674. These readings contrast with readings of 900 or above for a majority of our gonococci.

CONCLUSIONS

The CTA-carbohydrate media, the BACTEC® Neisseria differentiation system and the co-agglutination technique are highly effective for confirmatory identification of gonococci. Best results are obtained when the test is performed not from the primary culture but rather from a young subculture on chocolate agar. For the CTA and BACTEC® methods the subculture provides the heavy inoculum required for a rapid result. In addition, with the CTA method, results are obtained more rapidly when there is only a small amount of medium in the tubes. Although the BACTEC® method was 100% sensitive there was a problem in these studies with specificity for a few strains of Neisseria. False-positive results can, however, usually be recognized by an alert technologist because the glucose reading will be 600 or less, even when a heavy inoculum is used.

With the co-agglutination method, equivocal results were obtained for almost a third of the strains tested directly from antibiotic-containing medium. However, this problem was resolved when the test was repeated with growth from a subculture on chocolate agar. The need for subculture diminishes one of the primary advantages of the co-agglutination method, i.e., a rapid result obtained with only a few colonies of the organism on the primary culcute plate. Also, consistently satisfactory co-agglutination results were obtained only when micro-organisms were tested after boiling. Because half of our *N. lactamica* strains cross-reacted in this test, we recommend that an ONPG test be included when testing pharyngeal isolates. Since the ONPG reaction is usually complete by one hour, this does not significantly delay the culture report.

ACKNOWLEDGEMENTS

We extend our appreciation to Dr William Janda for providing us with cultures from the Howard Brown Memorial Clinic, to Johnston Laboratories for supplying BACTEC® Neisseria Differentiation kits, and to Pharmacia Diagnostics for providing Phadebact® Gonococcus Test kits.

REFERENCES

1. Handsfield, H.H. (1978): Gonorrhea and non-gonococcal urethritis. Recent advances. *Med. Clin. North Am., 62,* 925.
2. Kellogg, D.S. (1974): *Neisseria gonorrhoeae* (gonococcus). In: *Manual of Clinical Microbiology,* 2nd ed., Chapter 11, p. 124. Editors: E.H. Lennette, E.H. Spaulding and J.P. Truant. American Society for Microbiology, Washington.
3. Kellogg, D.S., Jr, Holmes, K.K. and Hill, G.A. (1976): Laboratory diagnosis of gonorrhea. *Cumitech* 4. American Society for Microbiology, Washington.
4. Strauss, R.R., Holderbach, J. and Friedman, H. (1978): Comparison of a radiometric procedure with conventional methods for identification of *Neisseria. J. Clin. Microbiol. 7,* 419.
5. Lind, I. (1975): Methodologic aspects of routine procedures for identification of *Neisseria gonorrhoeae* by immunofluorescence. *Ann. N.Y. Acad. Sci., 254,* 400.
6. Tronca, E., Handsfield, H.H., Wiesner, P.J. and Holmes, K.K. (1974): Demonstration of *Neisseria gonorrhoeae* with fluorescent antibody in patients with disseminated gonococcal infection. *J. Infect. Dis., 129,* 583.
7. Danielsson, D. and Kronvall, G. (1974): Slide agglutination method for serological identification of *Neisseria gonorrhoeae* with anti-gonococcal antibodies adsorbed to protein A-containing staphylococci. *Appl. Microbiol., 27,* 368.
8. Olcén, P., Danielsson, D. and Kjellander, J. (1978): Laboratory identification of pathogenic Neisseria with special regard to atypical strains: an evaluation of sugar degradation, immunofluorescence and co-agglutination tests. *Acta Pathol. Microbiol. Scand. Sect. B., 86,* 327.
9. Randall, E.L. (1975): Long-term evaluation of a system for radiometric detection of bacteremia. In: *Microbiology – 1975,* p. 39. Editor: D. Schlessinger. American Society for Microbiology, Washington
10. Kellogg, D.S., Jr, Peacock, W.L., Jr, Deacon, W.E., Brown, L. and Pirkle, C.I. (1963): *Neisseria gonorrhoeae.* I. Virulence linked to clonal variation. *J. Bacteriol., 85,* 1274.
11. Morello, J.A., Lerner, S.A. and Bohnhoff, M. (1976): Characteristics of atypical *Neisseria gonorrhoeae* from disseminated and localized infections. *Infect. Immun., 13,* 1510.
12. Catlin, B.W. (1975): Cellular elongation under the influence of antibacterial agents: way to differentiate coccobacilli from cocci. *J. Clin. Microbiol., 1,* 102.

Discussion

J. Maeland (Trondheim, Norway): After the suspension of bacteria has been boiled for the co-agglutination test, are the bacteria washed?

J. A. Morello: No. The suspension is simply boiled, a drop of the suspension then put on a slide together with a drop of the reagent, the two mixed and the reaction observed.

P. Olcén (Örebro, Sweden): We have also seen other cross-reactions in Örebro, namely with Haemophilus species. We think it is a good idea to do a Gram-stain in order to see the diplococci before doing the co-agglutination test.

J. A. Morello: In my presentation I omitted to mention that these tests on the clinical isolates were done only after an oxidase test and Gram-stains had been performed.

C. M. Anand (Calgary, Canada): How do these three tests compare in cost?

J. A. Morello: That is a good question. The most expensive test is the BACTEC®, which costs about $3 to perform, involving three vials. The ONPG (o-nitrophenyl-β-D-galactopyranoside) test must be set up separately. We prepared the CTA-carbohydrate media ourselves, but I believe that the commercially available ones cost about $1.50 per set. The co-agglutination test costs about $1.25 per test. The co-agglutination and CTA-carbohydrate tests are therefore similar, and the BACTEC® costs about twice as much as the other two. In addition, to do the BACTEC® test a BACTEC® instrument is needed.

Comparison of four techniques for the confirmatory identification of Neisseria gonorrhoeae

P. F. Jacobs*

Medical University of South Carolina, Charleston, South Carolina, U.S.A.

According to the United States of America National Morbidity and Mortality Report, gonorrhoea is still the most prevalent of the venereal diseases. There are a million new cases of gonorrhoea reported each year in the United States.

Historically, *Neisseria gonorrhoeae* has been identified using morphology and Gram reaction, indophenol oxidase activity and the fermentation pattern of certain carbohydrates in cystine trypticase agar (CTA) medium [1]. Once it has been established that the suspect organisms are oxidase-positive. Gram-negative diplococci, the carbohydrates glucose, maltose, sucrose and lactose are inoculated. The cultures should be incubated at 35 °C for up to 72 hours and results are often considered difficult to interpret because the minute amount of acid produced is often insufficient to change the phenol red pH indicator.

A modification of the CTA method was introduced by BBL Microbiologicals a few years ago [2]. The Minitek® system utilizes paper discs which have been impregnated with the carbohydrates dextrose, maltose and sucrose. Due to the fact that some organisms ferment lactose slowly, this carbohydrate has been replaced by a disc containing o-nitrophenyl-B-D-galactopyranoside (ONPG), which will detect the presence of β-galactosidase. This enzyme, which is produced by slow lactose-fermenting organisms, is responsible for cleaving the carbohydrate molecule. By the substitution of ONPG for lactose the test can be read in four hours.

Another definitive procedure available for identifying *N. gonorrhoeae* is a radiometric assay available through Johnston Laboratories for use with the Bactec® instrument [3]. This procedure is based on the measurement of liberated radiolabelled CO_2 from metabolized carbohydrates which have been tagged with ^{14}C. This assay uses glucose, maltose and fructose as its differentiating sugars, plus the ONPG reaction, and can be read in three hours.

Not all laboratories, e.g., Venereal Disease (V.D.) Control Laboratories and, perhaps, physicians' offices, need the ability to identify all of the Neisseria. For their purposes, the ability to identify *N. gonorrhoeae* definitively is sufficient.

Fluorescent antibody conjugate is available for both *N. gonorrhoeae* and *N. meningitidis* and is used in many areas as the method of choice for the definitive identification of these organisms, particularly if there is a discrepancy in the sugar

* Present address: Pharmacia Diagnostics AB, Uppsala, Sweden

fermentation results. This procedure usually takes several hours, most of which time is needed to allow the slides to air dry.

The present study was carried out to evaluate the Pharmacia Phadebact® Gonococcus Test. The principle of the Phadebact® Gonococcus Test is the co-agglutination of *N. gonorrhoeae* with IgG antigonococcal antibodies coupled by protein A bonds to heat-killed staphylococci. The Phadebact® Gonococcus Test is a two-minute slide co-agglutination procedure with an alternative boiling step recommended if results are questionable. The suspect organism is mixed into one drop of gonococcus reagent and is compared with the organism mixed into a drop of control reagent. The slide is rocked for two minutes and observed for agglutination.

A total of 116 cultures, 98 of which were obtained from a local V.D. clinic, were tested. The remaining 18 isolates were obtained from respiratory specimens submitted to the Medical University of South Carolina bacteriology laboratory and from quality control stock cultures. Each specimen was isolated on chocolate agar and was comprised of Gram-negative diplococci which were indophenol oxidase-positive. Each isolate was tested by Minitek®, Bactec®, fluorescent antibody staining (IFA), and the Phadebact® Gonococcus Test on the same day. Manufacturers' recommended procedures were followed in all cases. The alternative boiling step which is suggested when questionable results are obtained with the Phadebact® Gonococcus Test was not carried out due to insufficient growth and loss of organism viability.

The true positive or sensitivity rate and the true negative or specificity rate was calculated for each of the commercial systems tested. The Minitek® system identified 97 of 98 *N. gonorrhoeae* strains correctly, a true positive or sensitivity rate of 99%. All 18 other Neisseria species were identified correctly as non-gonococcal, a specificity of 100%. The Bactec® system identified all but one isolate of *N. gonorrhoeae* correctly for a true positive rate of 99% and identified 17 of 18 other Neisseria species correctly as non-gonococcal, for a true negative or specificity rate of 94.4%. The fluorescent antibody technique identified 93 of the 98 strains of *N. gonorrhoeae*, a sensitivity rate of 94.9%, and 17 of the other 18 Neisseria species as non-gonococcal for a specificity rate of 94.4%. The Phadebact® Gonococcus Test identified 96 of the 98 *N. gonorrhoeae* isolates correctly for a sensitivity rate of 98%. The two strains which yielded questionable results should have been retested after the alternative boiling step was carried out. All 18 of the other Neisseria species were correctly identified as non-gonococcal.

A comparison of some other parameters of the four systems shows that the fluorescent antibody technique requires about five minutes of a technician's time and about two hours to complete the entire procedure. Most of this time is spent allowing the slides to air dry. The capital equipment needed is a fluorescence microscope. The antisera cost $40.00 per vial or about $0.40 per test.

The Minitek® system requires about 15 minutes of technician's time and four hours of incubation. The equipment needed includes a humidor, pipetter and a disc dispenser. The cost of the broth, plates and discs is $52.50 or about $4.00 per test if quality control organisms are run with the unknown. This figure is based on one unknown plus three controls and will decrease substantially when several unknowns are tested at once.

The Bactec® system requires about 15 minutes of technician's time and three hours of incubation. Capital equipment investment is the Bactec® instrument. The kit containing the labelled carbohydrates and inoculating broth costs $30.00. A separate

ONPG system is $44.00, giving a cost per test of about $4.00.

The Phadebact® Gonococcus Test requires about three minutes of technician's time. No capital equipment is necessary. The test costs $50 or about $1.00–1.25 per test.

In conclusion, the Phadebact® Gonococcus Test appears to be reliable, requires a minimum amount of time, is easy to perform and its cost is competitive with that of other commercial methods.

REFERENCES

1. Kellogg, D.S., Jr (1974): *Neisseria gonorrhoeae* (Gonococcus). In: *Manual of Clinical Microbiology*, 2nd ed., p. 124. Editors: E.H. Lennette, E.H. Spaulding and J.P. Traunt. American Society for Microbiology, Washington, D.C.
2. Reddick, A. (1975): A simple carbohydrate fermentation test for the identification of pathogenic *Neisseria. J. Clin. Microbiol., 2,* 72.
3. Strauss, R.R., Holderbach, J. and Freedman, H. (1978): Comparison of a radiometric procedure with conventional methods for identification of *Neisseria. J. Clin. Microbiol., 7,* 419.

Comparison of immunofluorescence and different fermentation procedures for the identification of *Neisseria gonorrhoeae*

I. Moberg Kallings
Department of Bacteriology, National Bacteriological Laboratory, Stockholm, Sweden

During the sixties, with the steep increase in gonorrhoea incidence in Sweden, many attempts were made to improve the diagnosis and to standardize the diagnostic procedure throughout the country. The National Board of Health and Welfare made recommendations for the diagnosis and treatment of gonorrhoea in 1963, in the laboratories selective media such as Thayer-Martin were introduced and transportation was improved [1]

At the Neisseria Department of the National Bacteriological Laboratory (NBL), which is a reference as well as a diagnostic laboratory receiving 110,000 samples for gonococcal culture in 1978, immunofluorescence (IFL) was introduced 10 years ago. At that time a comparison was made between IFL and sugar degradation using the fermentation medium described by Juhlin [2], containing hemin, ascites and placenta. Testing of 1,071 strains produced an agreement of 99.9 %. Since then the medium for sugar degradation has been altered, and it was considered appropriate to make a re-evaluation of the procedures used for the verification of gonococci.

Of the cultured specimens 6.5 % are positive: in 1978, for example, there were 6,708 cultures to be verified from 4,348 patients, or an average of 25 suspect gonococci daily. As the amount of samples has an uneven distribution throughout the week, on some days there is a double number of strains to be verified.

MATERIAL AND METHODS

The diagnostic procedure used has been basically the same for 10 years. Specimens taken with a charcoaled cotton swab from urethra in men and from urethra, cervix and rectum in women are sent to the laboratory in a modified Stuart transport medium [1]. Each specimen is cultured on two media: one hematin agar (Difco peptone, Oxoid agar, horse blood and serum added) and one selective medium, e.g., the same hematin agar, additionally containing vancomycin 0.3 μg/ml and polymyxin B 25 IE/ml. The cultures are examined after 24 and 48 hours by stereomicroscope. Suspect colonies are tested for oxidase production, when positive a Gram stain is made, followed by IFL and subculture. With this procedure most cultures

Table I. Comparison of IFL and different sugar fermentation methods for verification of 210 strains suspected of being N. gonorrhoeae.

Number of strains	IFL	Methods for test of sugar fermentation			
		NBL	O-F	CTA	Elrod
Positive	202	202	Inconclusive results		
Negative	8	8			

can be pronounced positive two days after arrival at the laboratory. One positive culture from each patient is tested for antibiotic susceptibility by the disc diffusion method.

Suspect colonies from a throat specimen or from a rectum specimen, if that is the only specimen positive from one patient, are always tested for ability to attack glucose, maltose, fructose.

The reagents for IFL examination are prepared at NBL principally according to Danielsson [3]. Ten freshly isolated strains together with a representative set of auxotyped strains are used for the immunization of rabbits. The rabbit gonococcal antiserum is pepsin digested according to Forsum [4] to avoid cross-reaction with protein A of staphylococci. The conjugate is tested against a number of Neisseria species and the dilution to use is chosen so as to give no reaction with the apathogenic Neisseriae –1 : 20 is a commonly used dilution. The criteria to judge a gonococcal smear positive are: evenly distributed diplococci in at least three microscopic fields with a bright fluorescence of $3+ -4+$ round the entire edge of the bacteria.

The fermentation media used were one prepared at NBL with (Difco) base agar, horse serum, yeast extract and glutamin NAD, one CTA medium [5], one O-F medium [6], slightly modified, and Elröd-medium [7]. Cysteine-free medium of NEDA-composition [8] was used as a further verification of *Neisseria gonorrhoeae*.

Two hundred and two strains of *N. gonorrhoeae* of representative auxotypes [3] and eight commensal Neisseriae were tested in all four media and in IFL. Eleven *N. gonorrhoeae* that had created problems in identification and were sent to the NBL for typing and 19 *Neisseria meningitidis* that were 'weak' maltose-oxidizers were also tested. In addition, 571 consecutive clinical isolates from one month were tested in IFL and for fermentation in NBL medium.

Table II. Verification of N. gonorrhoeae by IFL. Specificity compared to biochemical method, e.g., sugar degradation using NBL medium.

Number of strains	IFL-positive		IFL-inconclusive, sugar degradation-positive		IFL-negative
	Day 1	Day 2	Day 1	Day 2	
571	476	54	70*	2	25

* Fourteen of these strains were never retested because the patient had isolates from other sites that were verified as gonococci

Table III. Verification of 532 N. gonorrhoeae strains by IFL. Duration for positive test as compared to sugar degradation.

		Day 1	Day 2
IFL-positive		89.5 %	99.6 %
Sugar degradation	glucose + maltose − fructose − }	−	91.6 %

RESULTS

Two hundred and two *N. gonorrhoeae* were positive in IFL and fermented glucose, but not maltose and fructose on NBL medium (Table I). They all grew well on the NBL medium. Of the eight other strains two were shown to be *Neisseria sicca*, three *Neisseria flava* and three *Branhamella catarrhalis*. They gave no or a very weak (< 1 +) reaction in IFL. All strains grew poorly or did not give a sufficiently distinctive reaction in the other methods used.

Of the 571 strains that were picked as suspect gonococci colonies on the examination of the culture plate by stereomicroscope and were shown to be oxidase-positive, 476 were positive in IFL on first examination (Table II). Fifty-four strains were positive in IFL after subculture, as well as being positive in the fermentation test on NBL medium. Two strains were still inconclusive in IFL on the second day but were positive in fermentation. Twenty-five strains were negative in IFL and on subculture no gonococci could be seen: these were pronounced negative. Fourteen strains were never retested as they grew poorly on subculture and strains from other sites of the same patient had been verified as gonococci.

Of the isolated *N. gonorrhoeae* 89.5 % could be pronounced positive the same day they were found on the culture plate when IFL is used for verification (Table III). Another 8.4 % could be decided on the next day and 0.4 % of the suspect strains needed still further examination.

The 11 strains of *N. gonorrhoeae* that had created problems in diagnosis were all positive in IFL, seven grew and fermented glucose on NBL medium on the first passage, four needed several passages to show a characteristic biochemical pattern, and none grew on cysteine-free medium (Table IV).

All 19 *N. meningitidis* strains showed a 2 + − 3 + 'spotted' appearance in IFL, 18 were glucose-positive, maltose-positive and fructose-negative in CTA and O-F medium, two strains did not ferment maltose on NBL medium and one strain did not ferment glucose in any of the media (Table V). They all grew on cysteine-free medium.

Table IV. Verification of 11 'problem strains' of N. gonorrhoeae by IFL and by sugar degradation.

	IFL-positive	Sugar degradation-positive (NBL)	Growth on cysteine-free medium
Number of strains	11	7	0

Table V. Sugar degradation of 19 strains of 'weak' maltose fermentors of N. gonorrhoeae.

	Method for test of sugar degradation			Growth on cysteine-free medium
	CTA	O-F	NBL	
Number of strains				19
Glucose +*				
Maltose +	18	18	16	
Fructose −				

* One strain did not ferment glucose in any of the methods used

DISCUSSION

IFL is a quicker method for verification of *N. gonorrhoeae* than fermentation procedures. It is almost as sensitive as the best of the fermentation methods used in our
comparison. It is well known, however, that IFL is not specific enough to differentiate
between *N. gonorrhoeae* and *N. meningitidis*. The reaction of the conjugate with
N. meningitidis cannot be completely eradicated by dilution, nor can the serum be
absorbed by meningococci without losing too many of the gonococcal antibodies.
Therefore, when *N. meningitidis* can be expected to occur more frequently (e.g.,
from throat cultures and from rectum cultures), fermentation procedures should be
adopted. Neisseria species other than gonococci are rarely found from urethral and
cervical specimens, in our material they were less than 0.05 %. Thus, for the diagnosis
of uncomplicated gonorrhoea, it is justified to use IFL for the verification of *N.
gonorrhoeae*.

As with all methods based on immunology, IFL is very dependent on the specificity
of available reagents and it has been difficult to find reliable commercial reagents.
It becomes a rather expensive and time-consuming method for the laboratory which
has to prepare its own reagents. IFL can never eliminate the need for biochemical
tests, but it is a good complement, especially for a laboratory with a large amount of
diagnostic specimens.

In search for a more defined fermentation medium than the hemin-ascites-placenta-
medium, several compositions without enrichment of biological material were
tried. Too many of our strains did not grow well enough or produce enough enzymes
until horse serum was added. This poor growth as well as the poor results when
using CTA, O-F and Elrod for sugar degradation tests, might be due to the high
percentage of poorly growing arginine-hypoxanthine-uracil-requiring strains in
this material [2].

As a very strong criterion for *N. gonorrhoeae*, the absence of growth on cysteine-
free media was used, as described by Catlin [8]. Of 800 *N. gonorrhoeae* tested on
cysteine-free medium none has grown while all of 76 *N. meningitidis* grew well.
It is an especially useful criterion when testing strains that have shown some problems
in differential diagnosis, e.g., *N. gonorrhoeae* that fail to ferment glucose or grow
too well on ordinary blood agar or *N. meningitidis* that fail to ferment maltose or
do so slowly or weakly.

REFERENCES

1. Gästrin, B. and Kallings, L.O. (1968): *Acta Pathol. Microbiol. Scand.*, *74*, 362.
2. Juhlin, I. (1963): *Acta Pathol. Microbiol. Scand.*, *58*, 51.
3. Danielsson, D. (1963): The demonstration of N. gonorrhoeae with the aid of fluorescent antibodies. *Acta Derm.-Venereol.*, *43*, 451.
4. Forsum, U. (1973): Characterization of FITC-labelled F (ab')$_2$ fragments of IgG and a rapid technique for the separation of optimally labelled fragments. *J. Immunol. Methods*, *2*, 183.
5. Moberg, I. (1975): Auxotyping of gonococcal isolates. In: *Genital Infections and Their Complications*, p. 271. Editors: D. Danielsson, I. Juhlin and P.-A. Mårdh. Almqvist & Wiksell, Uppsala.
6. Hugh, R. and Leifson, E. (1953): Taxonomic significance of fermentative versus oxidative metabolism of carbohydrates by various Gram negative bacteria. *J. Bacteriol.*, *66*, 24.
7. Catlin, B.W. (1974): Neisseria meningitidis. In: *Manual of Clinical Microbiology*, 2nd ed., chapter 10, p. 116. Editors: E. H. Lenette, E.H. Spaulding and J.P. Truant. American Society for Microbiology, Washington, D.C.
8. Catlin, B.W. and Carifo, K. (1973): *Appl. Microbiol.*, *26*, 223.

Practical evaluation of methods for the routine laboratory identification of *Neisseria gonorrhoeae*

F. Catalan and S. Levantis
Alfred Fournier Institute, Paris, France

INTRODUCTION

If the identification of *Neisseria gonorrhoeae* creates no problem for a specialized laboratory, it still constitutes a difficulty for medical laboratories, especially since the recrudescence of extragenital forms (cutaneous, pharyngeal and anal) of gonorrhoea.

The oxidase test is certainly very rapid, not expensive and easily available but is insufficient to establish that oxidase-positive bacteria belong to the genus Neisseria and still less to the type *N. gonorrhoeae*. Specific methods of identification such as the morphological appearance of colonies and sugar metabolism – which remains the reference method – followed by direct immunofluorescence (IF) with monospecific sera, are still very tedious, practicable for only a few laboratories.

In 1973, Kronvall's method of co-agglutination seemed to be very promising [1]. In a previous unpublished experiment in 1975, we demonstrated the interest of this method but the instability of the reagent caused some reservations. Later, a better-stabilized reagent was obtained which afforded the possibility of specific identification of *N. gonorrhoeae* and, especially, of its differentiation from other oxidase-positive bacteria with identical morphology.

The method is based on the principle of agglutination of a suspension of dead staphylococci the protein A of which is bound with specific anti-*N. gonorrhoeae* antibodies. A suspension containing staphylococci coated with IgG from a normal rabbit is used as a control.

The aim of the present study was to estimate the reliability and specificity of the Phadebact® Gonococcus Test in detecting *N. gonorrhoeae* among strains from patients.

The Gram-negative, oxidase-positive strains isolated were identified by the traditional sugar metabolism method. A comparative study was then performed between co-agglutination and indirect IF.

63

MATERIALS AND METHODS

Strains

Patient strains originated from patients attending the clinics at the St Louis Hospital and the Alfred Fournier Institute in Paris during the month of May, 1978.

Laboratory strains originated from lyophilized cultures in the Alfred Fournier Institute. All strains were sampled from the male urethra or pharynx.

Altogether, 110 bacterial strains were included in this investigation.

The identity of a strain was based on findings using the sugar metabolism method, taken as the reference method.

The origins and identities of strains are shown in Table I. They comprise:
– 78 strains of *N. gonorrhoeae* isolated from different samples,
– 20 strains of *N. gonorrhoeae* previously isolated and stored using lyophilization,
– 12 strains other than gonococcus, all oxidase-positive, isolated from different samples.

Cultures

Thayer and Martin medium [2] prepared in the laboratory of the Alfred Fournier Institute was used as growth medium for the primary isolates. The samples were cultivated for 18–24 hours at 35.5 °C, in an atmosphere of 10% CO_2.

Oxidase-positive colonies with typical morphology were tested in compartmented agar plates in Kellogg's medium containing a final concentration of 1% glucose, maltose, fructose and sucrose.

Immunofluorescence

Anti-gonococcal immune serum was raised in rabbits by intravenous inoculation

Table I. Bacteria tested.

Bacterium	Origin	Number of strains	
N. gonorrhoeae	Urethra	78	
	Laboratory strains	20	Total 98
N. meningitidis	Pharynx	2	
	Laboratory strains	4	Total 6
N. perflava	Laboratory strain	1	
N. mucosa	Laboratory strain	1	
N. subflava	Laboratory strain	1	
N. lactamica	Laboratory strain	1	
N. flava	Laboratory strain	1	Total 5
Moraxella	Laboratory strain	1	
Total		110	

64

of the F-62 strain of Kellogg. The antiserum obtained was absorbed against *Neisseria meningitidis*, *Neisseria perflava* and *Neisseria subflava*.

A sandwich technique was used.

The bacterial colonies to be tested were smeared out on a glass plate and fixed in acetone for two minutes. The smears were washed in a pH 7.2 phosphate buffer solution (PBS) and incubated with antigonococcus antibodies at 37 °C for 30 minutes, in a humid chamber. Non-bound antibodies were then washed with PBS and immuno-fluorescently labelled anti-rabbit IgG antibodies were added. The mixtures were incubated for a further 30 minutes as above and the excess of IF labelled antibodies was then rinsed away with PBS. The resulting immunofluorescent light pattern was inspected using light incident from the side. Fluorescence corresponding to 2+ or more was regarded as positive.

Co-agglutination

The reaction was performed on cultures from the primary plate and according to preliminary instructions for use. The reaction time was one to two minutes. For the freshly isolated strains, cultures 24 hours old were used; 36-hour-old cultures were used for the lyophilized strains.

The test was conducted in every case with the Phadebact® Gonococcus Test, a staphylococcal suspension coated with globulins from a non-immunized rabbit.

RESULTS

IF always gave good results with *N. gonorrhoeae* but also gave positive reactions with some strains of *N. meningitidis* (although most were 'marginal cases' with weak fluorescence = 2+) (Table II).

The co-agglutination technique gave better results than IF (Table III). All strains of *N. gonorrhoeae* were positive while strains other than *N. gonorrhoeae* never gave a positive reaction (Table IV). The intensity of the co-agglutination reaction was strong in 44% of cases, less strong in 56% but always clear compared to the control, which never gave a co-agglutination reaction (in contrast to the previous experiment in 1975).

Table II. Intensity of the IF reaction for six strains of N. meningitidis.

IF 3+ / Phadebact® negative	Group B wild strain
IF 2+ / Phadebact® negative	Autoagglutinable laboratory strain
IF 2+ / Phadebact® negative	Group A laboratory strain
IF 2+ / Phadebact® negative	Group B laboratory strain
IF 2+ / Phadebact® negative	Group B laboratory strain
IF 2+ / Phadebact® negative	Autoagglutinable laboratory strain

Table III. Comparison of results of Phadebact® and IF.

		Phadebact®		Totals
		+	−	
IF	+	98	6	104
	−	0	6	6
Totals		98	12	110

Table IV. Comparison of results of Phadebact® and sugar metabolism.

Bacterium	Number of strains	Phadebact® Gonococcus Test			
		Reaction	Reagent	Dextrose	Maltose
N. gonorrhoeae	98	Positive	Negative	+	−
N. meningitidis	6	Negative	Negative	+	+
N. perflava	1	Negative	Negative	+	+
N. mucosa	1	Negative	Negative	+	+
N. subflava	1	Negative	Negative	+	+
N. lactamica	1	Negative	Negative	+	+
N. flava	1	Negative	Negative	+	+
Moraxella	1	Negative	Negative	−	−

DISCUSSION

The co-agglutination reaction is generally clear and rapid, even though the strains correspond with the morphology described by Kellogg [3, 4]. Emulsification is no problem, while the control reagent gives an opalescent aspect very different to that obtained with the gonococcal reagent itself.

The most important step to be achieved is the preparation of the emulsion. The bacteria to be tested should be spread on the glass plate to a thin film before they are mixed with the reagent and control. In this way the suspension obtained is perfectly homogeneous and the test is easier to read. Both false positive reactions due to clumps of bacteria and false negative reactions due to too few bacteria reacting are avoided.

When the prepared reagent is too old the number of intensely positive reactions (4+) decreases rapidly. The majority of the reactions is weak (2+), although they remain distinct from the appearance obtained with the control.

The bacterial colonies should not be more than 24 hours old. On prolonged incubation the bacteria undergo autolysis and release material which interferes with the pH of the reagent and control.

On the other hand, it is important to emphasize that oxidase-positive bacteria other than *N. gonorrhoeae* never give a false positive reaction. Because of this the interest of the method is increased. IF sometimes gives reactions which are difficult to interpret except by a very experienced technician.

Apart from the remarks already made we find that co-agglutination proves to be more specific than IF with absorbed immune serum and to be equivalent to the biochemical method of sugar utilization. This being so, it is natural that the rapidity of execution of this technique makes it preferable to IF, the subjectivity of which is sometimes inconvenient.

In the case of acute male urethritis direct microscopic examination is sufficient, but it is different with the subacute disease where the organisms are extracellular and often accompanied by many other bacteria. Here, culture is indispensable but good judgement and long practice are necessary because the colonies are generally rare and the use of inhibitors is needed. In the case of subacute urethritis the use of the Phadebact® Gonococcus Test is, therefore, recommended.

The same situation exists when *N. gonorrhoeae* are isolated from extragenital samples. It is known that the oxidase-positive colonies of Gram-negative bacteria are not necessarily gonococci and only a biochemical identification (utilization of sugar in Kellogg's medium) allows diagnosis and differentiation of the saprophytic Neisseria. The Phadebact® Gonococcus Test reaction allows solution of the problem with one single colony, even if colonies are mixed with those of other bacteria of the commensal flora.

CONCLUSION

Co-agglutination using the Phadebact® Gonococcus Test proved to give 100% concordance with the reference method of sugar metabolism in the case of 98 strains of *N. gonorrhoeae*, six strains of *N. meningitidis*, five strains of saprophytic Neisseria and one strain of Moraxella.

IF gives similar results except for the strains of *N. meningitidis*, which occasionally give a fluorescence identical to that obtained with some strains of *N. gonorrhoeae*.

There is no difference between wild strains and lyophilized strains with either the Phadebact® Gonococcus Test reagent or with IF.

The results with co-agglutination are, therefore, in this investigation, as good as those with sugar metabolism but the method is simple, easy to perform and not very expensive, all the qualities needed for a routine test.

REFERENCES

1. Kronvall, G. (1973): A rapid slide agglutination method for typing pneumococci by means of specific antibody adsorbed to protein A-containing staphylococci. *J. Med. Microbiol., 6*, 187.
2. Thayer, J.D. and Martin, J.E. (1964): A selective medium for the cultivation of *Neisseria gonorrhoeae* and *Neisseria meningitidis*. *Publ. Hlth Rep., 79*, 49.
3. Kellogg, D.S., Jr, Peacock, N.L., Deacon, W.E., Brown, L. and Pirkle, C.I. (1973): *Neisseria gonorrhoeae*: I. Virulence genetically linked to clonal variation. *J. Bacteriol., 85*, 1274.
4. Kellogg, D.S., Jr, Cohen, I.R., Norins, L.C., Schroeter, A.L. and Reiseing, G. (1968): *Neisseria gonorrhoeae*: Colonial variation and pathogenicity during 35 months *in vitro*. *J. Bacteriol., 96*, 596.

Discussion

I. Lind (Copenhagen, Denmark): Were the culture suspensions boiled before the co-agglutination test?

S. Levantis: Not with the reagent described in this paper. Previously, with an older reagent, we were obliged to boil the suspension but now the reagent is better stabilized. As I think I said, we use a rather special technique in that there is a film layer over the glass slide and the agglutination is always clear.

I. Amirak (London, England): In a very short trial with the Phadebact® Gonococcus Test we found that, if the reagents are stable, they last a long time. We had two batches. The first gave very good results. As the expiry date approached we asked for another batch, but this did not give such good results. So we returned to the first batch which had by then passed its expiry date. We found it was quite stable and still gave extremely good results. It seems that the stability of the reagent is of great importance.

New test systems for the identification of *Neisseria gonorrhoeae*

J. S. Lewis
Department of Health, Education and Welfare, Public Health Service, Center for Disease Control, Bureau of Laboratories, Atlanta, U.S.A.

During 1977, gonorrhoea continued to rank first among reported communicable diseases in the United States. In a report entitled *Today's VD Control Problem* [1], the American Social Health Association, using Center for Disease Control (CDC) statistics, states that during the 12-month period ending December 31, 1978, United States gonorrhoea control programmes obtained culture specimens from some 8,600,000 women, of which 403,000 (4.7 %) were positive. Although the positivity rate was highest (19.5 %) in venereal disease clinics, only 10 % of all tests were performed at such clinics; 90 % of all tests were performed in other settings, such as private hospitals, family planning clinics and other health care providers. In these settings culture positivity rates in women ranged from 1.4 % in student health centres to 4.9 % for women in correctional centres. Among 1,900,000 women tested by private physicians, 35,500 (1.9 %) cultures were positive. The importance of routine, in-office, clinic, hospital and laboratory identification of gonorrhoea and its treatment is evident.

The signs and symptoms of gonorrhoea usually manifest themselves clearly within three to five days in infected men. In contrast, some 80 % of infected women have such negligible early symptoms that they fail to seek medical attention. This poses a dual problem. The asymptomatic woman is a hidden carrier of infection and, by the time she becomes aware of her symptoms, she may have developed serious problems such as the various complications of pelvic inflammatory disease. Here again is a need for the routine screening of sexually active women.

The CDC has published recommendations for the diagnosis of gonorrhoea for screening purposes [2].

To diagnose gonorrhoea in women the CDC recommends that separate culture specimens be obtained from the endocervical and anal canals. Diagnosis on the basis of clinical signs alone is not recommended in women. Gram staining or fluorescent antibody staining techniques are recommended only as adjuncts to culture.

To diagnose gonorrhoea in men, when Gram-negative intracellular diplococci cannot be identified in direct smears of urethral exudate, a specimen of urethral exudate should be cultured. In known or suspected homosexuals, an additional specimen should be obtained from the anal canal and pharynx.

Most laboratories associated with sexually transmitted disease (STD) clinics process large numbers of gonorrhoea cultures. Such laboratories have found that the most economical and reliable culture system is the use of selective media contained in standard 15×100 mm Petri dishes and incubated as soon as possible in a candle extinction jar. This system provides an 88–92 % sensitivity [3, 4] but cannot always be followed and may, in fact, not be best for small laboratories or the private physician's office. In these situations where fewer cultures are performed there are definite advantages to having media with a longer shelf life, incubation without a candle jar and *Neisseria gonorrhoeae* identification procedures which require less laboratory expertise. There are several major systems presently in use.

The commercially available products described below are all based on the presumptive identification of *N. gonorrhoeae*, i.e., on the finding of specimens from the genital tract yielding typical colonial growth of Gram-negative, oxidase-positive diplococci on modified Thayer-Martin or equivalent selective media.

Most of the diagnostic kits and systems are currently available in the United States. The evaluation of these kits has been continuous in the CDC Bureau of Laboratories for the past nine years. It has enabled the CDC to play a role in the critical assessment and improvement prior to marketing of a number of products developed in the United States and other countries.

All of these systems were compared with direct inoculation and incubation under the best possible conditions using strictly quality-controlled, fresh, selective medium. In most cases both fresh isolates and stock laboratory cultures have been used.

Isocult-GC®, manufactured by Smith Kline Diagnostics, consists of a plastic paddle which serves as the container for modified Thayer-Martin medium. The specimen is swabbed on the medium and a patented streaking device automatically streaks the specimen on the agar. The paddle is placed in a transparent flat-sided tube which serves as the isolation unit. A CO_2 generating tablet is then placed in the tube which is closed tightly. The unit is incubated in an upright position in a suitable incubator at 35 °C. After 24–48 hours of incubation, if any growth is apparent, oxidase reagent provided in the kit is applied to one side of the paddle. If positive colonies are observed, Gram staining is performed. If desired, the paddle may be sent to a confirming laboratory in a mailing carton which is also provided. Comparative studies yielded approximately 6–7 % fewer positive results using this system than with standard reference procedures [5, 6].

A system which has become very popular and has yielded results almost as good as the standard procedure is the JEMBEC® system, developed by John E. Martin of CDC in cooperation with the Ames Company. This is simply a rectangular plastic plate with a small well to accommodate a CO_2 generating tablet [7, 8]. JEMBEC® plates containing modified Thayer-Martin medium are inoculated, sealed in a zip-lock pouch and incubated. During incubation the humidity in the pouch rises causing the tablet to generate CO_2. This system is easy to work with since it needs less space than 100 mm Petri dishes, uses no candle jars and increases recovery of gonococci as compared with the Transgrow system. Loss of positives is in the 1–2 % range against the standard system but there are approximately 10 % more positives than with Transgrow [9].

The Gono-Pak® is simply a modification of the JEMBEC®/system using a 60 or 100 mm Petri dish and a CO_2 generating tablet in a larger zip-lock pouch. The major complaint with this test is that when the tablet generates CO_2 it spreads over the plate and is not aesthetically appealing to the laboratory worker. Again, there is approximately a 1–2 % loss of positives but this is much better than with Transgrow and the Gono-Pak ® is being used as a substitute for the latter.

The GC-Duet® and Gonorette® kits are manufactured by Bacti-Lab Inc. The GC-Duet® contains a two-section culture plate with modified Thayer-Martin medium. One side is for a cervical specimen, the other for a rectal specimen. The plates are supplied in a gas-impermeable heavy plastic film. Before sealing, the bag containing the plate is evacuated and the air replaced with pure nitrogen, creating an inert atmosphere in the bag. This product will not dry out or deteriorate for up to 90 days at room temperature or up to six months if kept refrigerated. The plates are of a slightly different design from the JEMBEC® plates, with hinges at the back.

Bio-Dynamics Inc. produce another kit called UNIBAC-GC®. This system provides everything necessary to perform a presumptive test for gonococcus and consists of six Transgrow bottles, swabs, glass slides, oxidase and Gram stain reagents.

Marion Laboratories manufacture a product containing a CO_2 tablet and an ampoule of dilute acid. The ampoule is crushed and the acid mixes with the tablet creating a CO_2 atmosphere equivalent to that in a candle extinction jar (approximately 3 % CO_2).

American Culture Media produce a combination system, the OB-GYN® plate, containing modified Thayer-Martin medium, Nickerson's agar for yeasts and eosin methylene blue (EMB) and blood agar as separate compartments.

Gonocult®, produced by Wampole Laboratories, is marketed only in Italy. It consists of a screw-capped disposable plastic tube containing modified Thayer-Martin medium. It is inoculated in the usual way. CO_2 from a pressurized can is sprayed into the tube for one second. Longer spraying would be detrimental to the gonococcus because 100 % CO_2 is used. Gases will not mix at the 10–90 air concentration. The product is inverted and incubated at 36 °C. The kit also includes oxidase reagents. A loss of approximately 10 % of positive cultures was obtained with this product as compared with the standard procedure, due to excessive CO_2 concentration and the small surface area of the medium [10].

Microcult-GC® is marketed by the Ames Company [11–14]. This system consists of a plastic slide containing two culture areas with modified Thayer-Martin medium in a dry reagent form. Rehydration of the culture areas is accomplished by the addition of seven drops of rehydration fluid (glycerinized water) just prior to inoculation with the specimen. The culture slide is sealed in the original foil pouch containing a CO_2 generating tablet and incubated. After incubation a cytochrome oxidase strip is pressed against the culture area. Any purple dots appearing on the culture area should be regarded as indicating the likely presence of gonococci. Smears should then be made from the positive oxidase locations and Gram stained to determine if Gram-negative diplococci are present. Tests are available in 10 culture slides per kit. The loss of positive cultures is approximately 3–5 % as compared with the standard procedure.

These kits for the presumptive identification of N. gonorrhoeae have all performed reasonably well. In general, 85–90 % sensitivity has been obtained, depending on the particular device. The kits offer good reproducibility, convenience of use, ease of storage and a saving of time through elimination of the need for preparing and sterilizing the necessary media. All the kits are more expensive than direct plating.

The present availability of these numerous new bacteriological systems suitable for physician's office or survey work should enhance the ability of various health care providers to screen diverse populations for gonorrhoea, to diagnose cases and improve follow-up of patients with recurrent infection.

REFERENCES

1. *Today's VD Control Problem* (1978). American Social Health Association, New York.
2. *Criteria and Techniques for the Diagnosis of Gonorrhea* (1975). Health, Education and Welfare, Public Health Service, Center for Disease Control, Atlanta.
3. Caldwell, J.G., Price, E.V., Pazin, G.J. et al. (1971): Sensitivity and reproducibility of Thayer-Martin culture medium in diagnosing gonorrhoea in women. *Am. J. Obstet. Gynecol.*, *109*, 463.
4. Schmale, J.D., Martin, J.E., Jr and Domescik, G. (1970): Observations on the culture diagnosis of gonorrhea in women. *Br. J. Vener. Dis.*, *46*, 201.
5. Lewis, J. S. (1970): unpublished data.
6. Kousa, M., Renkonen, O.-V., Kiistala, U. and Lassus, A. (1973): Transgrow (Clinicult) in the laboratory diagnosis of gonorrhea. *Br. J. Vener. Dis.*, *49*, 450.
7. Martin, J.E., Jr and Jackson, R.L. (1975): A biological environmental chamber for the culture of *Neisseria gonorrhoeae*. *J. Amer. Vener. Dis. Assoc.*, *2*, 28.
8. Martin, J.E., Armstrong, J.H. and Smith, P.B. (1974): A new system for the cultivation of *Neisseria gonorrhoeae*. *Appl. Microbiol.*, *27*, 802.
9. Martin, J.E. and Lester, A. (1971): Transgrow, a medium for transport and growth of *Neisseria gonorrhoeae* and *Neisseria meningitidis*. *Health Services and Mental Health Administration Health Reports*, *86*, 30.
10. Lazo-Wasem, E.A. (1975): Novel system for culture and transport of *N. gonorrhoeae*. *J. Am. Med. Technol.*, *37*, 213.
11. Lewis, J.S. (1977): Evaluation of a new gonorrhea culture detection system Microcult-GC. *Health Laboratory Science*, *14*, 22.
12. Sachs, G. and Hofherr, L. (1978): Evaluation of the Microcult-GC kit as a screening method for the detection of *Neisseria gonorrhoeae*. *J. Am. Med. Technol.*, *44*, 10.
13. Willcox, R.R. and John, J. (1976): Simplified method for the cultural diagnosis of gonorrhea. *Br. J. Vener. Dis.*, *52*, 256.
14. Unsworth, P.F., Talsania, H. and Phillips, I. (1979): An assessment of the Microcult-GC culture test. *Br. J. Vener. Dis.*, *55*, 1.

Discussion

I. Lind (Copenhagen, Denmark): Is there any problem with contaminated batches, e.g., with yeasts or moulds growing during incubation?

J. S. Lewis (Atlanta, U.S.A.): When these kits are produced commercially they are handled in about the same way as you would handle them. We have had contaminated batches, but for the most part it is not a problem.

An improved procedure for the Phadebact® Gonococcus Test

C. M. Anand

Provincial Laboratory of Public Health, Southern Branch, Calgary, Alberta, Canada

INTRODUCTION

A rapid slide agglutination test, utilizing the principle of co-agglutination, for serological identification of *Neisseria gonorrhoeae* was described by Danielsson and Kronvall in 1974 [1]. A commercial test kit based upon the co-agglutination technique was subsequently developed as the Phadebact® Gonococcus Test. The prototype and the kit have previously been evaluated and some of their shortcomings elucidated [2, 3]. The test kit was introduced in Canada at the beginning of this year with improved reagents. Because the Phadebact® Gonococcus Test promised to be a simple and quick technique for the confirmation of presumptively identified *N. gonorrhoeae*, the test was evaluated as a suitable replacement for or complement to the confirmatory procedures presently used in the Provincial Laboratories of Public Health in Alberta, Canada. These are the fluorescent antibody (FA) technique and the carbohydrate utilization test. The problems with and limitations of these procedures are well known [4–6].

Preliminary evaluation of the Phadebact® Gonococcus Test for identification of strains of gonococci isolated on Thayer-Martin medium according to the manufacturers' recommended procedure, however, gave a large number of inconclusive results and occasionally failed to identify gonococci. An alternative procedure, recommended for re-examination of strains giving non-interpretable results with the primary procedure, was, in fact, found to be more satisfactory for the initial examination of all strains, even though the time taken to elicit a strong reaction was found to be somewhat long — between seven and ten minutes. It was felt that this alternative procedure, modified to give a stronger reaction in a shorter time and at a reduced cost, would be a worthwhile test.

MATERIALS, METHODS AND RESULTS

A total of 448 strains of *N. gonorrhoeae*, 80 of *N. meningitidis*, seven of *Branhamella catarrhalis*, five of *N. lactamica* and two strains each of *N. subflava* and *N. sicca* were examined. The majority of these organisms were either 24- to 48-hour direct cultures or 24-hour subcultures of isolates from urogenital, rectal and pharyngeal specimens received at the Provincial Laboratories of Public Health in Calgary and Edmonton, Alberta, Canada.

Strains of *N. gonorrhoeae* were identified by the direct FA technique, using Difco reagents, and by the carbohydrate utilization test, using serum-free medium, de-

74

scribed by Flynn and Waitkins [7]. Other Neisseria species were identified by carbohydrate utilization reactions and additional bacteriological procedures as necessary [4, 8].

The reagents in the Phadebact® Gonococcus Test kits were reconstituted, the test and alternative procedures carried out, the reactions read and the results interpreted according to the package insert dated September, 1978.

Manufacturers' recommended procedure

The manufacturers' recommended procedure of testing colonies from culture medium directly on a microscope slide using the Phadebact® Gonococcus Test reagents was evaluated using three different media. This procedure is referred to hereafter as the 'standard' procedure.

Growths from the following media were examined:
1. Thayer-Martin (TM) medium [9] containing the following antibiotics: vancomycin (3 μg/ml), colistin (7 μg/ml), trimethoprim (3 μg/ml) and amphotericin B (5 μg/ml).
2. Imferon agar, a serum-free medium [10].
3. Edmonton Provincial Laboratory (EPL) medium, a serum-containing medium prepared from Tinsdale Base (Difco) and a supplement providing a final concentration in the medium of 0.71% osmotically lysed sheep erythrocytes, 8.7% bovine serum, 1.0% Isovitalex (BBL) and 0.39% disodium phosphate with antimicrobials as indicated for the Thayer-Martin medium, except that polymyxin B at a concentration of 2.4 μg/ml replaces colistin.

Selection of these media took into account the fact that TM medium is recommended by the manufacturer as giving best results and that there have been conflicting reports by previous investigators concerning the effect of serum-containing media on the results of co-agglutination [2, 3].

Growth on TM medium, even though only 24 hours old, was found to be adherent and sticky and could not be emulsified easily into the drops of reagent when the standard procedure was followed. EPL medium and Imferon agar were preferable

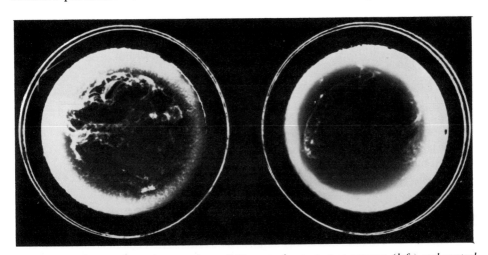

Fig. 1. Typical co-agglutination reactions of N. gonorrhoeae in test reagent (left) and control reagent (right) by the standard procedure.

Table I. Results of tests performed using manufacturers' recommended procedure on strains of N. gonorrhoeae grown on three media. (Percentages shown in parentheses.)

Test results	Edmonton Provincial Laboratory medium	Thayer-Martin medium	Imferon agar
Positive	20 (54)	19 (48.7)	36 (72)
Negative	1 (2.7)	2[a] (5.1)	10[c] (20)
Uninterpretable	16 (43.3)	18[b] (46.2)	4 (8)
Totals	37	39	50

[a] One strain was positive after five minutes
[b] Five strains were positive after five minutes
[c] Six strains were positive after five minutes

in this respect and Imferon agar was the better of the two. The reactions showed gluey strands in both the test and control reagents. The co-agglutination lattice tended to adhere to these strands making differentiation between the test and control reactions difficult when a less than strong reaction was obtained. This was particularly evident for strains isolated on EPL and TM media. Typical reactions are illustrated in Figure 1. There was no obvious difference between either of these two media in the number of positive versus negative reactions.

Of 39 gonococcal strains from TM medium, 19 gave clear positive results, two negative and 18 non-interpretable. Five of the 18 strains giving non-interpretable results, however, gave a clear-cut positive reaction after five minutes, as did one of the strains giving a negative reaction. From EPL medium, 20 out of 37 strains were positive, one was negative and 16 were non-interpretable. From Imferon agar, 36 out of the 50 strains were positive, only four were non-interpretable and 10 were negative. Six of these negative strains, however, gave clear positive results after five minutes.

The high proportion of non-interpretable results was found to be due to weak reactions with the test reagent when there was no reaction with the control reagent within the stipulated time limit or reaction in both test and control reagents.

In this investigation of the standard procedure, serum-free Imferon agar was the obvious choice both in terms of ease of performance and better results while serum-free TM and serum-containing EPL medium were equally unsatisfactory (Table I).

The new improved procedure

This is carried out as follows. Sterile distilled water (0.5 ml) is dispensed into 75 × 10 mm test tubes. Using a bacteriological loop, growth sufficient to give a dense suspension is removed from the culture plate and suspended in the water. A homogeneous suspension is then obtained by use of a vortex mixer. The suspension is held for five minutes in a boiling water bath and then centrifuged at 2,400 revolutions per minute (rpm) for five minutes. Using a Pasteur pipette, the supernatant is removed and three drops are returned to the deposit. The remainder is discarded and the cells resuspended. This suspension constitutes the antigen for the improved procedure.

Two circles are drawn with a china marker at either end of a dry, clean microscope

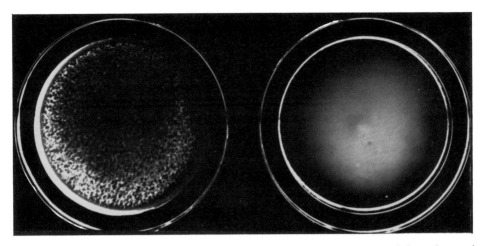

Fig. 2. Typical co-agglutination reaction of N. gonorrhoeae in the test reagent (left) and control reagent (right) by the improved procedure.

slide and labelled 'Test' and 'Control'. Using Pasteur pipettes, a drop (1/40 ml) of each of the test and control reagents is delivered to the respective circles on the slide. Similarly, one drop of the suspension is added to each circle. The slide is rocked gently and examined for co-agglutination using transillumination against a dark background. Co-agglutination in the test reagent and no reaction in the control reagent is considered positive for *N. gonorrhoeae*. Co-agglutination in the test and control reagents or no reaction in either is considered negative for *N. gonorrhoeae*.

The difficulty of emulsifying gonococcal growth in the reagent drops experienced in the standard procedure was eliminated with the boiled suspension. The vortexed suspension provided a clean suspension, free from strands. The speed and strength of the reaction were intensified and the appearance of co-agglutination with the test reagent contrasted well with the negative result in the control, making the interpretation of reactions much simpler. Typical reactions are illustrated in Figure 2.

Co-agglutination occurred within 60 seconds in most instances, although some strains required an additional 60 seconds and others up to five minutes. The alternative procedure described by the manufacturer gave weaker reactions and required longer time intervals for co-agglutination to occur. In view of these advantages the improved procedure was adopted as the best means for evaluation of the Phadebact® Gonococcus Test.

Of the 173 strains of *N. gonorrhoeae* grown on TM medium, 170 gave positive co-agglutination reactions for gonococci and three were negative. FA tests on these strains gave similar results although the three strains that were FA-negative are different from the three co-agglutination-negative strains. Of 176 strains from EPL medium, 175 gave positive reactions and one a negative. All 42 strains grown on Imferon agar gave positive reactions.

Seventy-nine out of 80 strains of *N. meningitidis* tested using the improved procedure showed no reaction in test or control reagent. One strain showed a minimal reaction in both. This reaction was interpreted as a negative result for *N. gonorrhoeae*. The strains of *N. lactamica*, *N. sicca* and *N. subflava* gave clearly negative results.

Table II. Reactions of different Neisseria species and Branhamella catarrhalis with the test and control reagents using the improved procedure.

Organisms	Co-agglutination reactions			Totals
	Test + Control −	Test − Control −	Test + Control +	
N. gonorrhoeae	387	4	0	391
N. meningitidis	0	79	1	80
B. catarrhalis	0	0	7	7
N. lactamica	0	5	0	5
N. subflava	0	2	0	2
N. sicca	0	2	0	2

Seven fresh isolates of *B. catarrhalis* showed co-agglutination in both the test and control reagents (Table II).

The use of a drop smaller than that from the reagent dropper bottles in the kit enabled 100–110 tests to be performed with each kit designed for only 40 tests. This provides a substantial saving in cost.

It has been observed that the speed and intensity of the reaction with the test reagent was related to the density of the suspension of the gonococcal cells. It was also obvious that one drop of reagent was being diluted with an equal amount of the suspension of gonococci. It was felt that if this dilution were eliminated by use of concentrated cells the time for co-agglutination by some of the strains of gonococci might be shortened. The effects of further concentration of gonococcal cells were therefore studied. The procedure previously described was followed to the completion of centrifugation. Using a Pasteur pipette, all of the supernatant was discarded and the deposit withdrawn and transferred in about equal amounts to the two areas of a clean slide marked 'Test' and 'Control'. One drop each of the test and control reagents was added to the appropriate area. The slide was rocked and results read as previously described. The intensity of reaction was increased and reaction time further reduced. Forty-two strains from Imferon agar and 43 strains from TM medium, including those previously giving slow or negative reactions, were tested. Results were obtained within 30 seconds in most instances. The strains that had previously given strong reactions after a period of delay now reacted within one minute. Three strains previously negative were still negative by this procedure.

DISCUSSION

The present investigation indicates that the test procedure recommended by the manufacturer does not bring out the full potential of the Phadebact® Gonococcus Test. A certain amount of expertise is required to read and interpret the results and an unacceptable number of inconclusive results is obtained. TM medium cannot be considered as the medium of choice. Results from serum-containing and serum-free media were similar.

Although a procedure suggested in the package insert as an alternative was better in this regard, worthwhile improvements have been made. In the new procedure, a homogeneous concentrated antigen is obtained by vortexing and centrifuging a heavy suspension of bacterial cells, while boiling time is reduced from 20 to five minutes. The strength and speed of reaction is increased and interpretation is simplified. The use of a smaller volume of reagent makes the test more attractive economically. Additionally, results are not affected by the medium used for cultural growth and the majority of gonococcal strains are reliably identified. Discrimination between *N. gonorrhoeae* and other Neisseria species is excellent.

The fluorescent antibody technique has one major advantage over the Phadebact® Gonococcus Test in that it can be performed from the direct plate when insufficient colonies are present to carry out the Phadebact® Gonococcus Test. However, the FA method has the disadvantage that fluorescent microscopy equipment, reagents and trained personnel are not within the scope of every laboratory. Additionally, the fluorescent antibody technique is not recommended for identifying Neisseria species isolated from the pharynx [5].

Other commonly used procedures for confirmation of *N. gonorrhoeae* are the carbohydrate utilization tests, problems with the preparation, use and interpretation of which are well recognized [4–7]. The Phadebact® Gonococcus Test has a distinct advantage over these procedures, particularly in terms of speed of results.

The Phadebact® Gonococcus Test performed according to the new method described here provides a simple means for rapid identification of *N. gonorrhoeae* that should have application both in laboratories dealing with large numbers of *N. gonorrhoeae* isolates and particularly in those handling smaller numbers, where carbohydrate utilization is often the only method available for confirmation. It has application in identifying isolates of 'typical colonies' from all sites, including the pharynx, from which *N. gonorrhoeae* may be isolated.

ACKNOWLEDGEMENTS

I am grateful to Mrs V.W. Kadis and Mrs M. Petley-Jones, both of the Provincial Laboratory of Public Health, Edmonton, for technical assistance and photographs, respectively. The Phadebact® Gonococcus Test kits were provided by Pharmacia (Canada) Ltd.

REFERENCES

1. Danielsson, D. and Kronvall, G. (1974): Slide agglutination method for the serological identification of *Neisseria gonorrhoeae* with anti-gonococcal antibodies absorbed to protein A-containing staphylococci. *Appl. Microbiol.*, *27*, 368.
2. Barnham, M. and Glynn, A.A. (1978): Identification of clinical isolates of *Neisseria gonorrhoeae* by a co-agglutination test. *J. Clin. Pathol.*, *31*, 189.
3. Menck, H. (1976): Identification of *Neisseria gonorrhoeae* in cultures from tonsillo-pharyngeal specimens by means of a slide co-agglutination test (Phadebact Gonococcus Test). *Acta Pathol. Microbiol. Scand. Sect. B*, *84*, 139.
4. Catlin, B.W. (1974): *Neisseria meningitidis* (Meningococcus). In: *Manual of Clinical Microbiology*, 2nd ed., Chapter 10, p. 118. Editors: E.H. Lennette, E.H. Spaulding and J.P. Traunt. American Society for Microbiology, Washington, D.C.
5. Washington, J.A., II, Martin, W.J. and Karlson, A.G. (1974): Identification of bacteria. In: *Laboratory Procedures in Clinical Microbiology*, 1st ed., Chapter 4. Editor: J.A. Washington II. Little, Brown and Co., Boston.

6. Shtibel, R. and Toma, S. (1978): *Neisseria gonorrhoeae*: evaluation of some methods for carbohydrate utilization. *Can. J. Microbiol., 24*, 177.

7. Flynn, J. and Waitkins, S.A. (1972): A serum free medium for testing fermentation reactions in *Neisseria gonorrhoeae. J. Clin. Pathol., 25*, 525.

8. Kellogg, D.S., Jr. (1974): *Neisseria gonorrhoeae* (Gonococcus). In: *Manual of Clinical Microbiology*, 2nd ed., Chapter 11, p. 127. Editors: E.H. Lennette, E.H. Spaulding and J.P. Traunt. American Society for Microbiology, Washington, D.C.

9. Thayer, J.D. and Martin, J.E., Jr. (1966): Improved medium selective for cultivation of *N. gonorrhoeae* and *N. meningitidis. Public Health Rep., 81*, 559.

10. Payne, S.M. and Finkelstein, R.A. (1977): Imferon agar: improved medium for isolation of pathogenic Neisseria. *J. Clin. Microbiol., 6*, 293.

Discussion

I. D. Amirak (London, England): What was the final pH and was there any variation? What was the support for the growth of the strains?

C. M. Anand: The Thayer-Martin medium is used in one laboratory and EPL medium in the other, in Alberta, the two laboratories being about 200 miles apart. Quality control of these media for routine use is done but I could not say what the pH is for each individual batch.

I. D. Amirak: It is just for personal interest because in some studies we have done, about which we will talk tomorrow, pH has affected the test to a certain degree. Was growth supported quite well on all three media?

C. M. Anand: Yes, on all three media. Insofar as the improved procedure is concerned, the medium was taken at random (whatever was being used at the time), the strains were routine isolates and the procedure was carried out to assess what would happen in a routine laboratory using such a kit. The results are based on this.

C. M. Woodmansea (Leeds, England): In all these cases with the three media, was a subculture from a primary plate being used?

C. M. Anand: Yes and no. If there was sufficient growth for the procedure to be carried out from the direct plate, that is what was done. If not, this was subcultured and the procedure done from that, so there was a mixture of direct cultures and subcultures.

J. R. J. Baenffer (Rotterdam, The Netherlands): Has any individual variation in results been found among the technicians who did the test? Our experience is that there is some difference in results depending on who does the test. If one technician does it and another repeats it, the same results are not always obtained. For example, we have reduced the volume of reagents and in the hands of one technician it did well but in the hands of another its performance was less satisfactory.

C. M. Anand: In the initial phase two of us did the test in our laboratory and two people in the Edmonton laboratory, who worked from Thayer-Martin. The problem we experienced was that it was very difficult reading the end-point using the recommmended procedure. The difference between, say, + and + + developed gradually. With the other procedure, however, the reaction obtained on the slide is so rapid that there is a good contrast with the control reagent and so there are no problems of variation between different technicians reading the results. I mentioned that with the standard procedure a certain amount of expertise is required before results can be reliably reproduced. This is not the case with the improved procedure.

Comparison between sugar degradation and serological tests for differentiation of pathogenic Neisseria

P. Olcén and D. Danielsson
Department of Clinical Bacteriology and Immunology, Central County Hospital, Örebro, Sweden

A seven-year-old boy, whose sister had rheumatoid arthritis, had been healthy before falling ill with high fever, headache and pain in the neck. On examination he had pharyngitis and minimal rigidity of the neck. Examination of the cerebrospinal fluid was normal. The symptoms waned without treatment but a relapse with high spiking fever occurred after a few days, with exanthema (maculopapular without pustules, some with petechiae) and arthralgia of the big joints. He was observed in hospital without treatment. The signs and symptoms faded but came back a few days later. This pattern was repeated several times. The diagnoses considered were drug fever due to salicylates, viral infection, subsepsis allergica and collagen disorder.

Oxidase-positive Gram-negative cocci were isolated from three out of seven blood cultures drawn on a routine basis. These cocci degraded glucose but no other sugar and disseminated gonococcal infection was diagnosed. Penicillin was given and the patient recovered.

An epidemiological investigation was conducted. No gonococci (GC) were isolated within the family and the parents were very upset by the diagnosis. The age of the patient and the lack of confirmatory epidemiological findings made the diagnosis of disseminated gonococcal infection undertain. Neither were the skin eruptions completely typical for this diagnosis. Characteristic skin eruptions due to gonococci are maculopapular, with pustules and without petechiae.

The bacterial strain was sent to us for serological examination. We identified it as a meningococcus, group B, type 2, without maltose-degrading capacity. In retrospect, the clinical findings were typical of a benign meningococcaemia, a diagnosis that was easier for the parents to accept [1].

Using only one diagnostic method such as sugar degradation (SD) for the diagnosis of GC and meningococci (MC) (Table I) may give false confidence as regards diagnostic accuracy. It is well known that atypical SD patterns occur [2] (Table II) but not how prevalent they are.

In the laboratory, using only SD for the diagnosis of pathogenic Neisseria, it may be possible to pick out some of the strains with atypical SD patterns when

Table I. Sugar degradation patterns of various Neisseria species.

Glucose	Maltose	Fructose	ONPG	Bacterium
+	−	−	−	GC
+	+	−	−	MC
+	+	−	+	N. lactamica
+	+	+	−	Other
−	−	−	−	Other

Table II. Sugar degradation patterns found atypically with known bacteria.

Bacterium	Glucose	Maltose	Fructose	ONPG
GC	−	−	−	−
MC	+	−	−	−
MC	−	+	−	−
N. subflava	+	+	−	−

SD diagnosis shows the site of isolation to be unusual, when the colonies look atypical or on the basis of clinical and epidemiological reasons.

The latter can be exemplified by another case report. The patient was a 59-year-old, previously healthy woman. Three days before admission to hospital she felt sick, vomited and had general fatigue without fever. Next day her arms and legs were painful and stiff and she had a low grade fever. On admission the patient could not move out of bed without help. The skin had petechiae on the right wrist and ankle. The right knee had signs of arthritis and direct microscopy of exudate revealed Gram-negative diplococci. There were no symptoms from the recto-urogenital (RUG) area and the patient had had no sexual contact in the previous 6 or 7 years.

Blood culture and culture of the joint fluid showed oxidase-positive, Gram-negative diplococci which only degraded glucose. The patient was treated with ampicillin for disseminated gonococcal infection and the signs of infection vanished. Serological investigations showed the strain to be a MC, group C, type 2. This gave a better explanation of the patient's disease.

There are many possibilities of supporting or discarding a diagnosis made by SD. Morphology apart, polysaccharide [3], protein [4] and lipopolysaccharide antigens can be examined by serological methods, chemical analyses relating to fatty acids and other structures can be performed and genetic studies for nucleotide percentage, transformation and deoxyribonucleic acid (DNA) homology may be used. The temperature, media and CO_2 requirements may be indicative and, in some cases, the antibiogram can be helpful. In the diagnosis of GC and MC serological methods seem to be the most practical additional diagnostic aids. The combination of different diagnostic principles is also common in other fields of bacteriology, for example in the diagnosis of Salmonella and anaerobes.

In order to compare SD with serological methods a study was undertaken in

Örebro [5]. During a ten-month period 97 isolates were randomly picked from RUG specimens and 77 isolates from throat specimens of oxidase-positive Gram-negative cocci with a colony morphology like that of GC or MC. Each isolate was subjected to co-agglutination (COA) [3], immunofluorescence studies (IFL) [6, 7] and examination of biological requirements. SD was performed with hemin-ascites-proteose (HAP) agar which includes human ascites [8] and using O-nitrophenyl-β-D-galactopyranoside (ONPG). The COA reagents were prepared from Cowan I staphylococci coated with absorbed rabbit antisera. One reagent covered the GC while four covered the MC groups A, C, D; B; X, Y, Z and 29E and W-135. The IFL study was performed with unabsorbed or, in some cases, with absorbed conjugates. A comparison of SD and COA diagnoses for the 97 RUG isolates is presented in Table III. Divergent results were seen in eight isolates.

One isolate from a rectal specimen was a MC group Z by COA but a GC by SD. Additional examinations with IFL and biological requirements as well as the morphology of the colonies supported the MC diagnosis.

Two isolates were negative in COA but GC by SD. Further examination supported the GC diagnosis and we conclude that these two isolates had antigenic structures not covered by our COA reagent.

Five isolates were GC by COA but neither GC nor MC in SD. Further examinations revealed that four of them, all isolated from the cervix, were contaminated with sugar-degrading bacteria and that pure cultures were GC in SD as well. One isolate did not degrade any of the sugars. Subcultures on glucose agar induced observable glucose degradation and the COA diagnosis was confirmed.

Table III. Comparison of SD and COA diagnoses for 97 RUG isolates.

		SD			Totals
		GC	MC	Neg	
	GC	75	0	5	80
COA	MC	1	0	0	1
	Neg	2	0	14	16
Totals		78	0	19	97

Table IV. Comparison of SD and COA diagnoses for 77 throat isolates.

		SD			Totals
		GC	MC	Neg	
	GC	12	9	0	12
COA	MC	3	33	0	36
	Neg	0	13	16	29
Totals		15	46	16	77

The four contaminated isolates apart, two isolates were correctly diagnosed with COA but not with SD and two isolates were correctly diagnosed with SD but not with our COA reagents.

A comparison of the SD and COA diagnoses in the 77 throat isolates is given in Table IV.

Thirteen of the MC diagnosed by SD were negative in COA and it is evident that more than 25 % of carrier MC were not groupable and thus not suited for serological diagnosis based on the presently accepted groups.

Three of the 15 isolates from throat specimens diagnosed as GC by SD were MC in COA (one group B, one 29E and one W-135) and on further evaluation. The number of isolates examined is small but points to a risk if only SD is used in the diagnosis of GC in throat specimens. As many as 20 % of isolates may be MC without maltose-degrading capacity.

Table V shows results with some strains sent to us from other laboratories here in Sweden. It is evident that the atypical SD pattern of MC is a reality. The IFL examinations using unabsorbed conjugates gave numerous cross-reactions between GC and MC but clearly differentiates these two species from others, with the exception of some *Neisseria lactamica* strains. Absorbed IFL conjugates can be used for

Table V. Comparison of SD, IFL, COA and morphological diagnoses for isolates from various geographical and anatomical sources.

Strain No.	Isolated from	Sugar degradation glu/mal/ fru/ONPG	IFL[a]		COA		Colony morph. like	Comments
			GC	MC	GC	MC		
1.	Blood	$-/+/-/-$	+	+	−	+(gr C)	MC	Probably MC C, unable to degrade glucose
2.	Throat	$+/-/-/-$	+	+	−	+(gr C)	MC	Probably MC C, unable to degrade maltose
3.	CSF	$+/\pm/-/-$	−*	+*	−	+(gr B)	MC	Probably MC B, weak maltose degrading capacity
4.	Blood	$-/+/-/-$	+	+	−	+(gr B)	MC	Probably MC B, unable to degrade glucose
5.	Blood	$\pm/\pm/-/-$	+	+	−	+(gr C)	MC	Probably MC C, weak fermenting capacity
6.	Throat	$-/+/-/-$	+	+	−	+(gr B)	MC	Probably MC B, unable to degrade glucose
7.	Blood	$+/-/-/-$	+	+	−	+(gr C)	MC	Probably MC C, unable to degrade maltose
8.	Nasoph.	$+/-/-/-$	+	+	−	+(gr C)	MC	Probably MC C, unable to degrade maltose
9.	Blood	$+/-/-/-$	+	+	−	+(gr B)	MC	Probably MC B, unable to degrade maltose

[a] IFL = immunofluorescence test with unabsorbed rabbit antisera (unless otherwise indicated)
* absorbed IFL conjugate

the differentiation of GC and MC but had a shelf life at $-20\,°C$ of just one month, which made them less useful.

Our conclusions concerning the diagnosis of GC are:

1) Urogenital isolates can be diagnosed with either SD or COA with about the same accuracy.

2) Nonurogenital isolates of suspected GC should be diagnosed by combining SD with other methods, for example COA [9]. The practice of 'one-armed diagnosis' for nonurogenital isolates of Neisseria is not recommended.

ACKNOWLEDGEMENTS

This work was supported by grants from the Swedish Medical Research Council (Project No. 4778), Örebro läns landsting and the World Health Organization.

REFERENCES

1. Olcén, P., Eeg-Olofsson, O., Frydén, A., Kernell, A. and Ånséhn, S. (1978): Benign meningococcemia in childhood. A report of five cases with clinical and diagnostic remarks. *Scand. J. Infect. Dis.*, *10*, 107.
2. Reyn, A. (1974): Family Neisseriaceae. In: *Bergey's Manual of Determinative Bacteriology*, 8th ed., p. 427. Editors: R.E. Buchanan and N.E. Gibbons. The Williams and Wilkins Company, Baltimore.
3. Olcén, P., Danielsson, D. and Kjellander, J. (1975): The use of protein A-containing staphylococci sensitized with anti-meningococcal antibodies for grouping *Neisseria meningitidis* and demonstration of meningococcal antigen in cerebrospinal fluid. *Acta Pathol. Microbiol. Scand. Sect. B*, *83*, 387.
4. Danielsson, D. and Olcén, P. (1979): Rapid serotyping of group A, B and C meningococci by rocket-line immunoelectrophoresis and co-agglutination. *J. Clin. Pathol.*, *32*, 136.
5. Olcén, P., Danielsson, D. and Kjellander, J. (1978): Laboratory identification of pathogenic *Neisseria* with special regard to atypical strains: an evaluation of sugar degradation, immunofluorescence and co-agglutination tests. *Acta Pathol. Microbiol. Scand. Sect. B*, *86*, 327.
6. Danielsson, D. (1963): The demonstration of *N. gonorrhoeae* with the aid of fluorescent antibodies. I. Immunological studies of anti-gonococcal sera and their fluorescein-labelled globulins, with particular regard to specificity. *Acta Derm. Venereol.*, *43*, 451.
7. Danielsson, D. and Forsum, U. (1975): Diagnosis of Neisseria infections by defined immunofluorescence. Methodologic aspects and applications. *Ann. N. Y. Acad. Sci.*, *254*, 339.
8. Juhlin, I. (1963): A new fermentation medium for *N. gonorrhoeae*, HAP-medium. Influence of different constituents on growth and indicator colour. *Acta Pathol. Microbiol. Scand.*, *58*, 51.
9. Danielsson, D., Olcén, P. and Sandström, E. (1977): Serological methods of diagnosis. In: *Gonorrhoeae. Epidemiology and pathogenesis*, p. 27. Editors: F.A. Skinner, P.D. Walker and H. Smith. Academic Press, London, New York, San Francisco.

Discussion

J. A. Morello (Chicago, U.S.A.): Were the maltose-negative meningococci sulphadi-azine-resistant?

P. Olcén: I do not know. Do you think there is a correlation between sulphadiazine resistance and sugar permeability or degradation capacity?

J. A. Morello: Did you try subculturing them several times?

P. Olcén: Yes, we did. It is rather interesting. We have wondered about the menin-gococcal strains that are glucose-negative and maltose-positive. Why is it that they cannot utilize glucose but can use maltose? I should like to ask whether anyone has any suggestions.

J. A. Morello: We had one strain like that in the study I presented. The only way we were able to pick up the positive glucose was with the Bactec® system. It would be interesting to try some of these strains with it. The system seems to be the most sensitive in picking up the very small degradations of carbohydrate that other carbohydrate systems cannot. That is all I can say.

J. Henrichsen (Copenhagen, Denmark): I was slightly confused by the use of the word 'probably' in Dr. Olcén's last Table. If you only dare to say 'probably', why not do some fermentation assays?
  Secondly, did you try to heat-treat the bacteria used for absorption?

P. Olcén: The bacteria are heat-treated.

J. Maeland (Trondheim, Norway): Did Dr Olcén do all his fermentation testing on the same medium? I think it is common experience that results may vary from one medium to another. If all the fermentation testing was done on the medium de-scribed by Juhlin I feel sure there would have been results different from those ob-tained if, for example, the serum-free medium described by Flynn and Waitkins had been used.

P. Olcén: Concerning the medium described by Juhlin, it is quite clear that it is hard to get human ascites. It has to be from a patient who does not have diabetes and who has received no antibiotics or cytostatics. We previously obtained several litres of ascites, storing it deep-frozen. The ascites used in these studies was from the same patient. It is correct, as Dr Maeland suggested, that ascites is different from different patients at different times and that each ascites has to be tested by the laboratory using a range of bacterial strains before being accepted. I also agree with Dr Maeland that media with different compositions will give different results but even today we do not know which of the media gives the best results.

J. Maeland: Our experience in the laboratory is that we very frequently had trouble with fermentation testing using the medium described by Juhlin. When we changed to a different medium, about 50% or more of the problems were eliminated and results were much more clear-cut.

Summary, conclusions and general discussion

In the Chair: I. Lind (Copenhagen, Denmark)

I. Lind: I would like to open the discussion by presenting the results obtained with Phadebact® Gonococcus Test in our laboratory.

Our results are very similar to those described by Dr Morello and partly similar to those obtained by Dr Anand. In the preliminary experiments we used the procedure recommended by the manufacturer (Table I). The essential finding was that the test was negative for nine out of 25 strains of *Neisseria gonorrhoeae* and inconclusive for five strains. The meningococci caused no problems. One strain of *Neisseria lactamica* out of two gave a typical positive reaction.

Table I. Results obtained with Phadebact® Gonococcus Test performed according to the recommended procedure.

Species	Number of strains	Phadebact® Gonococcus Test		
		Positive	Negative	Non-interpretable
N. gonorrhoeae	25	11	9	5
N. meningitidis	8	0	8	0
N. lactamica	2	1	1	0

The experiments were, therefore, repeated using the alternative procedure, i.e., boiling a suspension of the bacteria for 10–20 minutes (Table II). Only gonococcal strains which were negative using the recommended procedure were retested. Eight of the nine previously negative strains now gave a typical positive reaction. We decided, in consequence, to use the alternative procedure in subsequent experiments. In order to obtain sufficient material for a dense suspension of bacteria, subcultures were always prepared. It is very seldom that the primary culture gives a sufficient amount of growth to ensure clear-cut, rapid co-agglutination.

Table II. Preliminary results obtained with Phadebact® Gonococcus Test performed according to the alternative procedure.

Species	Number of strains	Phadebact® Gonococcus Test		
		Positive	Negative	Non-interpretable
N. gonorrhoeae	9	8	1	0
N. meningitidis	3	0	3	0
N. lactamica	3	0	3	0

Table III. Examination of 230 strains belonging to Neisseria species with Phadebact® Gonococcus Test performed according to the alternative procedure.

Species	Number of strains	Phadebact® Gonococcus Test		
		Positive	Negative	Non-interpretable
N. gonorrhoeae:				
urogenital isolates	146	136	7	3
pharyngeal isolates	9	6	2	1
N. meningitidis	46	0	46	0
N. lactamica	8	3	5	0
Other Neisseria species	22	0	8	14

The investigation (Table III) comprised 155 gonococcal strains, 146 of which were urogenital isolates and the remaining nine pharyngeal isolates. Seventy-six strains belonging to other Neisseria species were tested, including 45 of *N. meningitidis* and nine of *N. lactamica*. In accordance with the protocol, all gonococcal strains were identified both by the direct immunofluorescence test and by bacteriological methods, which means Gram-stained smears and carbohydrate utilization tests.

Phadebact® Gonococcus Test correctly identified 93 % of the urogenital isolates. False-positive reactions occurred in three out of eight strains of *N. lactamica*. The O-nitrophenyl-β-D-galactopyranoside (ONPG) test, which is rapid and easy, might, therefore, be useful as a supplement to the co-agglutination test, if used on pharyngeal specimens.

I should like to ask Dr Rudin how the rabbit antiserum used for the co-agglutination reagent is produced, how many gonococcal strains are used, from whence they have been collected and what is the immunization procedure. Since the way in which the reagents are prepared is extremely important, I should like to ask, in particular, how many strains are used for immunizing the rabbits and how these strains are selected.

L. Rudin (Pharmacia Uppsala, Sweden): With regard to the source of collection of the strains, when research and production were started we had strains collected only from Sweden. However, it was soon realised that reagents prepared using these strains did not identify strains from other parts of the world, so we started to collect from the United States, also, and some from Canada and Germany. The United States strains were obtained from the Center for Disease Control, from Dr Johnston in Dallas, and in two cases from the West Coast, from Dr Falkow. Those from Canada were from Dr Diena's group.

I cannot say how many strains are used, not because there is something I wish to hide but because I do not know the exact number. As far as I remember, it is between 20 and 25. The strains have been chosen on the basis of our own experience and also of the serogrouping done by Dr Johnston.

I. Lind: Are you referring to serogrouping based on the outer membrane complex?

L. Rudin: Yes. We think that the outer membrane proteins are those which are important in preparing the antisera. During a prolonged period of research different combinations and pools of these strains have been tried. They have now been gathered into four or five pools which are injected successively into the rabbits. The immunization procedure is essentially that published by Danielsson.

I. Lind: Does that mean by intravenous injections of the antigen?

L. Rudin: Intramuscular injection with Freund's adjuvant, first, and then intravenous injections with the pooled strains. After the first round of intravenous injections the rabbits are bled to determine whether there is a sufficiently high titre. If not, the procedure is repeated until the titre in the co-agglutination test is high enough.

I. Lind: Do you think that various auxotypes and antimicrobial susceptibility patterns are represented among these strains?

L. Rudin: I do not know, because those aspects have not been considered. We have simply taken the strains that gave good results according to our own experience. We have also taken advice from other people.

I. Lind: The auxotypes vary from one geographical area to another. The strains isolated in Scandinavia, for instance, are very different from those isolated in many other places.

L. Rudin: That is our experience, also.

J. S. Lewis (Atlanta, U.S.A.): From information presented today and from our own work, I would suggest that Pharmacia should not change anything it is doing with regard to the strains being used and the immunization schedule. Sensitivity has not been a problem but rather, it seems to me, the methodology.

I. Lind: It is strange, however, that there is a need to use boiled suspensions in some laboratories whereas in other laboratories the test works well using a direct procedure. I thought that the reason might be found in a discrepancy between the antigens of the gonococcal population present in that area and the antibodies used for the reagent.

G. L. Daguet (Paris, France): From an academic point of view, I would like to know whether any species of Neisseria occur naturally in the rabbit.

I. Lind: If sera from old rabbits are tested in a gonococcal complement fixation test more than 50 % give a positive reaction. It was shown many years ago by Dr Reyn that this cross-reactivity is due to the presence of Pasteurella species in the noses of the rabbits. I do not think any Neisseria species were detected.

L. Rudin: Our experience is that sera from old rabbits (non-immunized animals) will contain antibodies that co-agglutinate Neisseria species. When control reagents are prepared, therefore, sera from very young rabbits must be used.

I. Lind: Yes. We have the same problem when we produce gonococcal antisera for use in the direct immunofluorescence test. We use rabbits weighing about 2 kg for immunization, i.e., very young rabbits, and they are selected by means of the gonococcal complement fixation test to ensure that they do not have cross-reacting antibodies.

Table X on page 8 relates to identification procedures. We would probably all agree that, in most cases, a confirmatory diagnosis of the isolate is necessary. Today we have discussed and learnt about the following three tests:
1. the carbohydrate utilization test and its various forms,
2. the direct immunofluorescence test and
3. the co-agglutination test.

Our experience with the carbohydrate utilization tests has been similar to Dr Moberg's. We have stopped using the medium described by Juhlin which requires the use of ascitic fluid. For some months we used the serum-free medium described by Flynn and Waitkins but too many gonococcal strains were unable to grow on it. Subsequently, we developed another medium which is enriched by the addition of fresh, sterile guinea-pig serum. This medium is expensive but very efficient. All strains grow to an extent which ensures a distinct colour change. Has anyone tried something of this kind, replacing ascitic fluid with an alternative? I imagine other people must have had problems using ascitic fluid in their media.

P. Olcén (Örebro, Sweden): We have tried the medium discussed by Dr Moberg. We have not investigated it particularly but from time to time the horse serum contained maltase which ruins the test because it breaks down maltose to glucose and fructose, with the result that a gonococcus will be misdiagnosed as a meningococcus. After discussing this problem, it seems that this happened purely by chance. It was one horse which sometimes had maltase in its serum and sometimes did not. Has Dr Lind used human serum instead of guinea-pig serum?

I. Lind: Yes, we have tried horse, pig, sheep, rabbit, guinea-pig and human sera. Difficulties resulting from the presence of maltase in the sera occurred with horses, pigs and sheep. Some batches of human serum inhibited the growth of nearly all the gonococcal strains and we therefore do not use the human serum. Rabbit serum is quite good but does not support the growth of the more delicate strains as well as fresh guinea-pig serum. If the guinea-pig serum is heat-inactivated, it is not fully efficient in supporting the gonococcal growth.

J. A. Morello (Chicago, U.S.A.): With regard to the problems everyone has in getting gonococci to grow, I wonder whether the cystine-tryptic digest agar (CTA) system which I described today is of interest. In fact, this is not the one we use in our laboratory – we use a system which has an agar base, increasing the carbohydrate concentration to 2 % and again using a very heavy inoculum of organism. We in fact ran this system parallel with the study I presented today and results have been reported elsewhere. We found that, depending not on growth but just on preformed enzymes, the results with this system approached those with BACTEC®, with almost 100 % identification of all live strains. This system simply depends on the enzymes already in the organism and does not try to include anything which would allow them to grow. I wonder whether there is something that I do not know about

Neisseria, or the medium, since most people here seem to feel that they should have a growth system rather than a system which detects preformed enzymes for the degradation of the carbohydrate. This system has been very successful for us and gives very rapid answers, usually within one or two hours.

I. Lind: The reason we use a growth-dependent method is that the technicians prefer it.

P. Olcén: If preformed enzyme is to be detected, perhaps that enzyme has to be induced previously. The media from which the bacterial batch is taken can be critical. Must it be taken from the same medium or from any of the different sugars? Is glucose degraded only if glucose has been available in the medium?

J. A. Morello: We never try to make the identification using growth from primary plates. We always subculture to chocolate agar, obtaining a full plate of growth and then using the entire growth from the plate to inoculate all the carbohydrates. We have had a very few strains which will not give a positive reaction using these conditions.

P. Olcén: The explanation for the negative reactions may be permeability difficulties. Perhaps the outer membrane should be sonicated or disrupted, leading to even better results as regards detection of preformed enzyme.

J. A. Morello: For 99 % or more of the strains we have in our laboratory there is no such problem.

S. A. Waitkins (Liverpool, England): Being partly responsible for the Flynn-Waitkins medium, I suppose I must defend it somehow! I accept that there are sometimes equivocal results with the serum-free media. In the United Kingdom we tend to inoculate the medium heavily, otherwise the bacteria will not grow on the serum-free medium. It is also important to be very sure that the final pH is 7.2. Also, as has been mentioned, guinea-pig serum is difficult for most of us to obtain, but the commerical supplements can be simply picked off the shelf as the media batches are made up. We have found the serum-free medium reasonably satisfactory.

I. Lind: Do you know whether there are many AHU-(arginine-hypoxanthine-uracil)-requiring strains in the United Kingdom?

S. A. Waitkins: That is a problem. We do not usually auxotype the strains and it is quite possible that such strains would cause difficulties.

I. Lind: They are predominant in Denmark.

H. Arvilommi (Jyväskylä, Finland): Dr Morello is not the only person not very knowledgeable about the growth of Neisseria. We had many problems with the growth-dependent fermentation test and tried several methods without much success.

J. A. Morello: I agree. With respect to the AHU strains, at least as regards the CTA medium, it is my understanding that it contains no uracil so that AHU strains will not grow, yet we are consistently able to obtain positive reactions using a large inoculum of AHU strains. Even if the strain is not auxotyped, the AHU strain can be recognized because it produces a smaller colony. Usually, if we say something will be an AHU because of colony size, the auxotyping result will turn out correspondingly.

J. S. Lewis: I am interested to know how many people here are using Phadebact® Gonococcus Test and whether they have had any additional problems?

M. Barnham (Harrogate, England): I looked at the prototype reagents about three years ago. Our findings were rather similar to those from Canada. There was no significant difference using serum-containing and serum-free media. Both gave very similar reactions. We performed a small experiment to determine whether there was much technician variation in reading a set of results. The major difficulty at that stage was that the control reagent was giving positive reactions and that it was impossible to distinguish these from reactions obtained with the gonococcal reagent. The reagents were subsequently improved, the control became clearer and we found that the various technicians who observed the reactions interpreted them very similarly. From doing quite a large number of these tests my feeling was that one improved as time went on. I do not know whether this is still true because I do not use this test routinely at the moment although I will probably take it up again. In those days I did not boil the strains, but we obtained fairly clear-cut reactions even with the unboiled reagents.

L. Rudin: It should perhaps be mentioned that the directions for use have been rewritten. We now recommend the boiling procedure as first choice, and say the results are stronger and easier to interpret. We still have the two procedures, the direct colony and the boiling procedures, as alternatives.

I. Lind: I think that problems with the identification of cultures from serum-containing medium existed only at the very beginning, when the control reagent consisted of naked staphylococcal cells. Later, the control cells were coated with IgG from non-immunized rabbits.

A. A. Lindberg (Stockholm, Sweden): I have to testify that I am also a non-expert on gonorrhoea. The statement that it is most likely the gonococci and their protein antigens which are being detected puzzles me, however. I think when you asked about the immunization procedure you were talking about the outer membrane complex proteins.

I. Lind: No. My question to Dr Rudin was whether he could characterize the strains he has used. I do not think it is protein antigens that are reacting.

I. Henrichsen (Copenhagen, Denmark): Instead of using these carbohydrate fermentation tests, has anyone ever tried to look at the possibility of working with monospecific direct enzyme tests, such as the ONPG test? There are various other possibilities, too.

I. Lind: There are a few papers describing studies in which people have tried to characterize Neisseria species by their enzymatic profiles but I have not seen any reports on the use of these characteristics for diagnostic purposes.

J. Henrichsen: It seems to work so well within other bacterial groups.

J. A. Morello: The API group has been able to develop a system to characterize Neisseria by enzymatic profiles but I am not certain which enzymes are included. There is a publication in the *Journal of Clinical Microbiology* about a year ago describing this system, but the system is not yet available for testing. I tried to get hold of it.

J. Maeland (Trondheim, Norway): We have been talking a great deal about co-agglutination tests but we should also mention the slide-agglutination test that has been described by Diena and colleagues and which has proved to be very useful in the identification of isolates of gonococci. The antibody that works is directed against lipopolysaccharides (LPS) and not against the protein antigens of the outer membranes.

I. Lind: Have you tried the test?

J. Maeland: I have, but not in exactly the way it was described by Diena and colleagues – not for identification but for other purposes.

I. Lind: We obtained some of the hen anti-LPS serum from the WHO, but neither of the procedures described works in our hands. There is a recent modification of the test, which we have not tried, in which the addition of nuclease to the test system is recommended. We found the same degree of agglutination or clumping with the control reagent and the anti-LPS serum. I do not know why.

I will now try to draw a conclusion from our discussions on the identification procedures. The conclusions for a large routine laboratory are different from those for a smaller laboratory. I think that the direct immunofluorescence test is the method of choice for the large laboratory. Among the methods discussed today it is the only one by which it is possible to make a diagnosis from primary cultures for 85–90 % of the isolates. This means that the result can be obtained within 24 hours after receipt of the specimen.

On the other hand, the method is difficult to handle if the laboratory is not large enough to produce its own reagents and to maintain technical expertise. For smaller laboratories it seems that a choice must be made between a carbohydrate utilization test and the co-agglutination test. The co-agglutination test gave a correct confirmatory diagnosis in about 95 % of cases in all laboratories, as far as I remember, which seems satisfactory.

Does anyone else have conclusions they wish to draw from today's presentations or from the information we have been given?

J. R. J. Baenffer (Rotterdam, the Netherlands): I would wish to make an exception for the pharyngeal strains. Do you prefer to use the immunofluorescence method for them too?

I. Lind: No. Several methods are necessary and the best practice is to combine an immunological test with some of the bacteriological identification procedures. We use the same procedure as that described by Dr Moberg.

J. Maeland: Dr Lind showed that the co-agglutination reagent, Phadebact®, did not give a positive reaction with meningococci. I wonder whether that is possibly because the antisera used for the reagent are absorbed with meningococci?

I. Lind: As far as I know they are. Perhaps Dr Rudin could discuss that?

L. Rudin: That is the critical point, the point at which most of the antisera are destroyed. We have to absorb out the cross-reactions, thereby also decreasing the specific titre.

I. Lind: The meningococci can, in fact, be identified when the suspension of bacteria is boiled because of the characteristic clumping that occurs.

J. S. Lewis: Why do you not use absorbed antisera?

I. Lind: I have never used absorbed antisera for the direct immunofluorescence test. If absorption is used, reactivity with about 15 % of the gonococcal isolates is lost.

J. S. Lewis: That would then eliminate the need to run another procedure to do the throat specimens.

I. Lind: It is not possible to absorb out the cross-reactivity with meningococci without losing the reactivity with a certain amount of the gonococcal strains. I do not think that anybody has been able to find a meningococcal strain removing the cross-reactivity without also removing gonococcal antibodies.

II
Beta-haemolytic streptococci

The status of developed and developing techniques for rapid group identification of streptococci

R. R. Facklam

Clinical Bacteriology Branch, Bacteriology Division, Bureau of Laboratories, Center for Disease Control, Public Health Service, Department of Health, Education and Welfare, Atlanta, U.S.A.

INTRODUCTION

Since the introduction of serological grouping procedures by Rebecca Lancefield in 1933 many modifications of the original procedure have been developed. Most of these modifications have been directed towards simplifying the laboratory procedures. The original procedure and the modifications all require about the same amount of time for final identification: about two days after the specimen has been obtained from the patient. One technique, the fluorescent antibody (FA) technique, was developed in the 1960s. This technique could be used to identify group A streptococci within four to five hours after obtaining the specimen from the patient. There is no doubt that the FA technique is very useful, especially in laboratories where hundreds of thousands of throat swabs are processed yearly. The FA technique, however, does have several disadvantages compared to the newly developed grouping procedures. These disadvantages will be discussed later.

Compared to the conventional grouping technique, the more recently developed techniques for serological grouping, such as co-agglutination, latex agglutination and nitrous acid extraction, shorten the time required for identification by one full day. Where health care is based on identification of the pathogen causing illness, early identification of the pathogen is desirable. The identification procedure should also be simple and inexpensive in order to maintain accuracy and efficiency. The conventional grouping procedures are too time-consuming and are often subject to error because of their complexity. The development of simple, rapid and inexpensive laboratory techniques for the identification of streptococci is, therefore, welcome.

CURRENT STATUS

I would now like to report to you what I believe is the current status of these newly developed techniques. I will report on our findings at the Center for Disease Control (CDC) as well as other published results.

Table I shows the time, complexity and cost requirements for four rapid identification techniques. The three newly developed techniques (nitrous acid extraction, co-

99

Table I. Rapid identification procedures for streptococci.

Procedure	Requirements		Complexity
	Time	Cost/test	
Nitrous acid extraction	30 minutes	$ 1.00	Complex
Fluorescent antibody	4–5 hours	$ 1.00	Complex
Co-agglutination	16–20 hours	$ 2.00–3.00	Simple
Latex agglutination	16–20 hours	$ 2.00–3.00	Complex

agglutination and latex agglutination) must be compared to the established rapid technique (the FA procedure). The FA procedure is complex but relatively inexpensive after a sophisticated microscope is purchased. The nitrous acid extraction procedure of throat scraping requires the least amount of time to complete and is the least expensive. However, it is comparatively complex and the process of throat scraping with a plastic spatula is distressing to the patient. To my knowledge this technique has been used in only one laboratory. It should, therefore, be considered experimental at this point. The two new agglutination procedures require about the same amount of time to complete and cost about the same. The co-agglutination procedure is less complex because it does not require an extract, as does the latex agglutination procedure.

Table II lists the different serological reagents that are available for each of the rapid techniques. Commercially-available precipitating antisera are used with the nitrous acid extraction procedure. The micronitrous acid extraction procedure works very well with the β-haemolytic group A, B, C, F and G streptococci. It does not extract the group D streptococci very well (most of these are not β-haemolytic). FA reagents are available for the six most common group antigens of streptococci, A, B, C, D, F and G. However, because of lot-to-lot variation in cross-reactions, only the reagent for group A streptococci is reliable. Experiments in our laboratory by Dr H. Wilkinson have demonstrated that good group B FA reagent can be prepared but no commerical company has made the reagent according to this procedure.

Table II. Streptococcal reagents available for rapid grouping techniques.

Nitrous acid extraction	A, B, C, D, F, G [1]
Fluorescent antibody	A, (B, C, D, F, G) [2]
Co-agglutination	A, B, C, G
Latex agglutination	A, B, C, D, F, G

[1] Commercially-available precipitating antisera
[2] Commercially available but unreliable

Table III. Results and needs of nitrous acid, fluorescent antibody and latex agglutination grouping of β-haemolytic streptococci.

Procedure	Antigen preparation			
	Direct from throats	Four-hour broth	Extraction of:	
			16-hour plate	40-hour broth
Fluorescent antibody	NA	95 %	–	–
Nitrous acid	96 %	NA	100 %	100 %
Latex agglutination	NA	NA	100 %	100 %

NA = data not available.

The latex agglutination reagents are more complete than the co-agglutination reagents. Reagents for all six groups, A, B, C, D, F and G, are available. The co-agglutination reagents currently available are for groups A, B, C and G. The manufacturer is developing a group D reagent that will be available in 1980.

Table III gives some of the published results of using FA nitrous acid extraction and latex agglutination procedures for identifying β-haemolytic streptococci. The FA procedure cannot be used for identifying group A streptococci from swabs taken directly from the patient's throat. The swab must be incubated in broth for four or more hours and then tested for the presence or absence of group A streptococci. This procedure identifies 95 % of the group A streptococci. Extracts are not used with the FA procedure but the latter can be used to identify group A streptococci by using the growth from blood agar plates (data not shown). The nitrous acid extraction procedure as alluded to earlier can be used to identify β-haemolytic group A, B, C, F and G streptococci directly from the throat, or from as few as three or four colonies from blood agar plate growth or from broth cultures. Ninety-six per cent of the β-haemolytic group A streptococci were correctly identified by

Table IV. Results and needs of co-agglutination grouping of β-haemolytic streptococci.

Study	Antigen preparation				
	Direct from plate	Four-hour culture		16-hour culture	
		cells	super-natant	cells	super-natant
CDC	NA	90/100 %	NA	90/100 %	100 %
Others	93 %	85 %	97 %	99 %	NA

NA = data not available.

extracting throat scraping with nitrous acid extraction techniques. There is perfect correlation between the nitrous acid extraction procedures of growth taken from blood agar plates and growth in broth and conventional techniques. Group D streptococci are excluded from the studies with the nitrous acid extraction procedure. No data are available for the four-hour broth technique. The latex agglutination technique has not been used to detect β-haemolytic streptococci directly from throat swabs or from four-hour broth cultures. There is 100 % correlation between latex agglutination and conventional grouping procedures if the extracts of β-haemolytic streptococci are made from a 16-hour blood agar plate growth or from an overnight broth culture (40 hours removed from the patient).

Table IV lists the results obtained in the CDC streptococcus laboratory and the results obtained by other investigators using the co-agglutination procedure. Antigens can be prepared in several ways. Firstly, cells can be taken directly from the blood agar plate and reacted with the co-agglutination reagents. Secondly, four-hour broth culture suspensions or culture supernatants can be used. Thirdly, overnight (16-hour) broth culture suspensions or culture supernatants can be used.

The results are all quite similar. We did not test the co-agglutination reagents using the direct method. We have had conflicting reports on its usefulness. However, at least one investigator reported a 93 % accuracy in identifying β-haemolytic streptococci by this procedure. The percentages listed as 90/100 under four-hour and 16-hour culture cell suspensions refer to the results obtained when we modified the cell suspensions with trypsin. Ninety per cent of the specimens were identified correctly without trypsin, but the remaining 10 % had to be trypsinized to remove the multiple reactions that cause indecisive results. Two drops of a 5 % solution of trypsin were added to each suspension that gave multiple reactions. This mixture was incubated at 37 °C for 30 to 60 minutes before retesting. When this procedure was used, all the strains were correctly identified. The best results with the co-agglutination reagents were obtained when the broth supernatants were used as antigens. Other investigators have reported a 97 % correlation between co-agglutination and conventional grouping procedures with four-hour broth supernatants of β-haemolytic streptococci. Although we have not used four-hour broth supernatants, we have tested 16-hour broth supernatants with the co-agglutination reagents. We observed a 99 % correlation between the co-agglutination and conventional grouping procedures.

In summary, we can conclude that all three newly developed rapid grouping techniques (nitrous acid extraction, latex agglutination and co-agglutination) are more convenient and less expensive than conventional techniques. The reliability of the nitrous acid extraction procedure needs to be established, for there is only one report on evaluation of the technique. The reliability of the latex agglutination grouping procedure also needs to be established. However, the results in our laboratory indicate that the technique works very well with β-haemolytic streptococci. The reliability of the co-agglutination grouping procedure is well established. There have been several published reports indicating a very high degree of correlation between the co-agglutination and conventional grouping procedures.

There are several limitations to these procedures that should be heeded. None of the three procedures have worked well with group D streptococci. The nitrous acid technique does not extract group D antigen. Although the latex reagents contain a group D reagent, not all group D streptococci are identified by the reagent.

If group D strains are accidentally reacted with the co-agglutination reagents, they may be misidentified as group A, B, C or G streptococci. Thus group D streptococci must be identified by another procedure.

The manufacturers' instructions should be followed precisely when using the latex agglutination or co-agglutination reagents. Modified procedures should be carefully evaluated and compared with established procedures before adopting them for routine use in the clinical laboratories. When these procedures are carefully monitored with quality control strains, they should be useful for all clinical microbiology laboratories.

REFERENCES

1. Arvilommi, H. (1976): Grouping of beta-hemolytic streptococci by using coagglutination, precipitation, or bacitracin sensitivity. *Acta Pathol. Microbiol. Scand. Sect. B, 84*, 79.
2. Arvilommi, H., Uurasmaa, O. and Nurkkala, A. (1978): Rapid identification of group A, B, C and G beta-hemolytic streptococci by a modification of the co-agglutination technique. Comparison of results obtained by co-agglutination, fluorescent antibody test, counter-immunoelectrophoresis and precipitin technique. *Acta Pathol. Microbiol. Scand. Sect. B, 86*, 107.
3. Christensen, P., Kahlmeter, G., Jonsson, S. and Kronvall, G. (1973): New method for the serological grouping of streptococci with specific antibodies absorbed to protein A-containing staphylococci. *Infect. Immun., 7*, 881.
4. El Kholy, A., Facklam, R., Sabri, G. and Rotta, J. (1978): Serological identification of group A streptococci from throat scrapings before culture. *J. Clin. Microbiol., 8*, 725.
5. Facklam, R., Cooksey, R.C. and Wortham, E.C. (1979): The evaluation of commercial latex-agglutination reagents for grouping streptococci. *J. Clin. Microbiol., 10*, in press.
6. Hahn, G. and Nyberg, I. (1976): Identification of streptococcal groups A, B, C and G by slide coagglutination of antibody-sensitized protein A-containing staphylococci. *J. Clin. Microbiol., 4*, 99.
7. Lue, Y.A., Hewit, I.P. and Ellner, P.D. (1978): Rapid grouping of beta-hemolytic streptococci by latex agglutination. *J. Clin. Microbiol., 8*, 326.
8. Moody, M.D., Ellis, E.C. and Updyke, E.L. (1958): Staining bacterial smears with fluorescent antibody. IV. Grouping streptococci with fluorescent antibody. *J. Bacteriol., 75*, 553.
9. Rosner, R. (1977): Laboratory evaluation of a rapid four-hour serological grouping of groups A, B, C, and G streptococci by the Phadebact streptococcus test. *J. Clin. Microbiol., 6*, 23.
10. Slifkin, M., Engwall, C. and Pouchet, G.R. (1978): Direct-plate serological grouping of beta-hemolytic streptococci from primary isolation plates with the Phadebact streptococcus test. *J. Clin. Microbiol., 7*, 356.
11. Stoner, R.A. (1978): Bacitracin and coagglutination for grouping of beta-hemolytic streptococci. *J. Clin. Microbiol., 7*, 463.

Discussion

I. D. Amirak (London, England): In the costing exercise, $ 2 to $ 3 was given for a couple of the methods but the cost for the nitrous acid extraction method was said to be more than $ 1 – how much more?

R. R. Facklam: It should have said *less* than $ 1. It is a very inexpensive method. All that is required basically is the nitrous acid, the glacial acetic acid and the commercially-available antisera. It costs less than $ 1, the exact amount depending on what has to be paid for the precipitating antisera.

W. R. Maxted (London, England): Dr Facklam is evidently very pleased with all these methods and his 100 % results but he does not mention any cross-reactions with any of them, which I do not believe.

   With regard to the list of sera, for example latex with that group D serum, it works very well for *Streptococcus faecalis* and so on, but not for all of the group D streptococci, such as *Streptococcus bovis*, as is well-known. What can be said, therefore, about the reliability? Do people have to use their own judgement on any of these tests because of cross-reactions? It all looks too good to be true.

   Thirdly, I know that Dr Facklam tried nitrous acid and that the Egyptians make it work very well, but our experience with it has been very unhappy. Perhaps someone else from another reference laboratory has tried it. I would be interested to hear about it.

R. R. Facklam: First, referring to the cross-reactions, Dr Maxted and I may have something of a difference in philosophy here about grouping. On these slides I was referring to β-haemolytic streptococci. If we open the grouping procedures to non-β-haemolytic streptococci, the situation is different. I will agree exactly with what he said. If there is a group D reagent present, that group D reagent should be expected to react monotypically with the group D streptococci, but Dr Maxted is right, it does not.

W. R. Maxted: It is not only group D but some of the other serum preparations which give cross-reactions and I am talking about β-haemolytic streptococci. I think everybody has had the experience of the group C reacting with almost everything.

R. R. Facklam: With which reagent?

W. R. Maxted: Certainly with Phadebact® reagent, depending on the broth used, and to some extent with the latex also, all with the β-haemolytic streptococci.

R. R. Facklam: That may be right, but the package insert states simply that Difco® Todd-Hewitt broth should not be used, which is the broth that is causing the problem. I will not dispute the fact that cross-reactions can and do occur but I am saying

104

that those cross-reactions can be resolved. One thing that can be said about the Wellcome reagents – the Streptex® or latex kits – is that there are no cross-reactions among the β-haemolytic streptococci. I have tested over 200 strains of various groups. The Wellcome reagents are monotypic, following their protocol and advice. The only problem I had with the Wellcome reagents is their group D reagent. With the non-β-haemolytic streptococci, following their procedure of extractions from plates, 75 % of the *S. bovis* strains are probably missed because they will not react. If the alternative procedure is used, that is, growing the strain up in a 20 ml broth and doing a pronase extract, there will be cross-reactions. Viridans streptococci will cross-react with the group D reagent. This leads to misidentification. Basically, the group D reagent is not very good, but that does not negate the usefulness of that product for identifying β-haemolytic streptococci.

There were several problems with Phadebact® reagents when they first came out and I did not obtain any better than 50–60 % correct identification. But these problems were basically resolved by modifying the procedure. We eliminated the broth reaction by changing the broth to make the antigen. When a broth other than Difco® Todd-Hewitt was used, most of the cross-reactions were eliminated.

I do not think I live in Utopia either but the package insert also states that the faster and stronger reaction is the one to read – and the technicians get used to doing this. It is much easier than looking through a microscope for nine hours a day.

W. R. Maxted: I still do not agree. I think that these reagents contain cross-reacting antibodies.

J. Rotta (Prague, Czechoslovakia): The nitrous acid extraction is definitely a very impressive method. My name is on one of the papers which has just been published. I wanted to repeat the experiments described by the Cairo laboratory. We have simulated the situation by picking one, two or three colonies and running a nitrous acid extraction. We have not done very many experiments so far but, for some reason, we have been unable to repeat the method. We cannot obtain a clear-cut precipitation reaction using the nitrous acid extraction procedure. I have no explanation for this. I wonder whether the throat scraping technique would be better. I still think that the technique will require further investigation.

R. R. Facklam: That was my criticism of the nitrous acid extraction technique. There has only been the one official report. I have used the micronitrous acid extraction technique in my laboratory with about 60 stock strains of cultures which are used for the evaluation of grouping reagents. It worked all the time but it was rather difficult to adjust the pH because we were working with microlitre amounts of material, 20 μl of nitrous acid and 10 μl of glacial acetic acid. It is a nice clean, clear-cut method, and the precipitant reactions I had were monotypic. We reacted them with all our available grouping antisera but I used CDC® grouping antisera which are substantially more free of cross-reactions than most commercial products. However, I would expect Professor Rotta's sera to be as good as mine.

M. Slifkin (Pittsburgh, U.S.A.): I want to make a few comments about direct tests, especially co-agglutination methods. Phadebact® procedure permits 24-hour and four-hour agglutination procedures. A question then arises about a direct procedure.

If we are given a pure culture of an organism on a plate, a direct co-agglutination procedure can be followed without difficulty. But, as some people have already said, what happens when we want to do a co-agglutination procedure from a primary plate? I believe it can be done, but certain protocols have to be followed. At least from my own observations, a minimum of 10 colonies is necessary. If there are fewer than that, the chance of getting a co-agglutination response with Phadebact® reagent, or perhaps any co-agglutination response, is lessened. The more colonies there are the better the chance of getting a response.

Working with primary plates there can be difficulties if, along with the β-haemolytic organisms, there are other organisms picked up, such as staphylococci and viridans streptococci, certainly leading to cross-reactivity. I think that the problems other people have observed with the direct procedure from primary plates have probably been associated with not having enough organisms or, concomitantly, having other organisms there which cross-react.

R. R. Facklam: With anything new that people start to do, I think that they tend to expect things to come out exactly the way described. Certainly, Dr Slifkin found that he was able to identify 95 % of his strains. Invariably, however, the basic techniques have to be learned. The little intricacies about what has to be avoided and what has to be looked for, in order to work through the various complications of a new technique that may be encountered, must also be learned. Nothing is as clean and clear-cut as we would like it to be. Conventional techniques are totally unacceptable for clinical laboratories that need an immediate answer. It is all right for reference laboratories like mine and we will never give up our conventional techniques of doing a Lancefield extract and precipitin test, but clinical laboratories need something which will provide an answer rapidly.

Addendum by Dr Facklam

A thought occurred to me after the symposium was over. I believe I may be able to explain why Dr Maxted's and Professor Rotta's experiments were unsuccessful with the micronitrous acid extraction techniques. The precipitin reaction between the nitrous acid extract and the grouping antisera is not the same as the conventional extracts, such as the Lancefield or Fuller methods. During the extraction period with nitrous acid sodium acetate is formed. When the nitrous acid extract is reacted with the grouping antisera the sodium acetate causes a 'salting-out' effect of the immune complexes, in this case the reactions between the streptococcal antigens and antibodies. The reaction becomes diffused throughout the capillary within a few minutes and stronger reactions form clumps of precipitates throughout the capillary tube. I do not think the ring or modified ring precipitin test can be used with the nitrous acid extract. We use the modified ring precipitin tests with conventional extracts and we observe a visible precipitate at the interface between the two reagents. This does not occur with the nitrous acid extracts. The reaction is diffuse throughout both reactants.

The role of microbiological examination in the diagnosis and prevention of streptococcal diseases

J. Rotta

Institute of Hygiene and Epidemiology, Prague, Czechoslovakia

Streptococcal diseases represent a *world-wide* health and economic problem. Available data on the incidence of streptococcal infections indicate that they are one of the most frequent bacterial diseases of man in the temperate zone and are common in tropical and subtropical regions.

A number of important clinical, microbiological and epidemiological observations have been made which have led to considerable improvement in the diagnosis, therapy and, partly, prevention of these diseases. In spite of these contributions, morbidity from streptococcal infections continues to be relatively high.

The following data document the *size of the problem.*

Prospective studies in various countries have disclosed that in *moderate climatic areas* as many as 20 % or more of individuals may be harbouring haemolytic streptococci in some situations. Data reported in recent years from *tropical and subtropical* countries indicate that the carriership there is often no less [1–4].

Table I. Incidence of upper respiratory tract infections. (Nine-year study; 4,000 subjects.)

	Nasopharyngitis	Pharyngitis	Laryngitis
Incidence per 1000 subjects/year	93	84	18
Streptococcal aetiology (%)	4.8	43.0	6.5

Table II. Accuracy of clinical diagnosis of pharyngitis. (One-year study; 24,300 subjects.)

		Pharyngitis	
		Streptococcal	Non-streptococcal
No. of cases	Clinically diagnosed	938	989
	Bacteriologically confirmed (%)	490 (52.5)	646 (65.5)

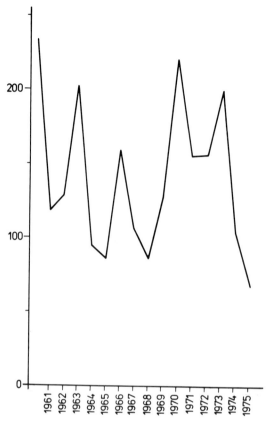

Incidence of scarlet fever, 1961–1975, among a 100,000-strong population group.

The incidence of *streptococcal acute upper respiratory tract disease* in the *moderate zone* is 5–15 cases per 100 individuals per year [5].

Our nine-year prospective study [6], completed in 1975 in a sample of the general population amounting to 4,000 inhabitants, showed that almost half of the pharyngitis cases were of streptococcal aetiology but that streptococci played only a very small role in nasopharyngitis and laryngitis (Table I).

Data on scarlet fever followed over a period of 15 years in the same geographical area but covering a 100,000-strong population group indicated a variation in incidence from some 70 cases to 240 cases per year (Fig.).

While the diagnosis of scarlet fever can be adequately made on the basis of clinical symptoms, the diagnosis of streptococcal pharyngitis based on *clinical parameters only* is highly unreliable. For example, in a one-year study of ours in a population of 25,000, clinically-postulated streptococcal pharyngitis could be bacteriologically confirmed in 52.5% of cases and non-streptococcal pharyngitis in 65.5% of cases (Table II).

This and similar reports from other countries clearly indicate that *bacteriological examination* of all cases of pharyngitis is necessary for *reliable diagnosis*. The practical consequences with regard to penicillin therapy, protection of patients against sequelae and control of spread of streptococcal infection on the one hand and the avoidance of unnecessary penicillin treatment on the other are obvious.

In *hot climatic areas* streptococci tend to produce rather mild respiratory disorders, the frequency of which has not yet been adequately estimated. However, skin infections due to haemolytic streptococci are very common there and their prevalence can even reach 20% or more of the child population at particular seasons [7–10].

The incidence of *rheumatic fever* has declined in the developed countries of the moderate zone but seems to have stabilized at the low figures reached. In most countries of Europe and in the United States of America the incidence amounts to some five attacks per 100,000 per year (Table III) [11, 12].

In the developing countries in tropical and subtropical regions, in contrast, rheumatic fever and rheumatic heart disease represent a health problem of roughly the same magnitude as used to be the case in countries of the moderate zone several decades ago (Table IV) [13, 14].

The *requirement for laboratory services* carrying out microbiological examinations in different areas of the streptococcal field stems from the following facts:

a) The group A streptococcus is a major cause of bacterial upper respiratory tract diseases, the accurate diagnosis of which is never certain if based on clinical examination only.

b) Acute infection may lead to rheumatic fever and acute glomerulonephritis only if it has been caused by a group A streptococcus.

Table III. Incidence of rheumatic fever in some European countries.

Country	Number of cases per 100,000	Date of study	Population size (millions)	Age group (years)	Comments
Denmark	29.5	1952	4.0		Compulsory
	15.5	1960	4.5	all ages	notification (1)
	10.7	1970	4.9		
England & Wales	15.8	1959	1.2		Compulsory
	8.8	1961	1.3	1–14	notification (2)
	4.7	1963	1.3		
Norway	72	1946			Compulsory
	28	1956	4.0	all ages	notification (3)
	14	1964			
The Netherlands	19	1959–1965	0.13	all ages	Prospective study (4)
Czechoslovakia	61.2/42.9	1961	0.1/0.05		
	22.8/19.3	1966	0.3/0.07	< 15/ > 15	Prospective study (5)
	8.5/2.6	1972	0.3/0.08		

Sources of data: (1) National Health Service
(2) Ministry of Health
(3) Central Bureau of Statistics
(4) Reference [11]
(5) Reference [12]

Table IV. Rheumatic fever (RF) and rheumatic heart disease (RHD) in some countries outside Europe.

Country	RF incidence per 100,000	RHD prevalence per 1,000	Date of study	Population size (thousands)	Age groups (years)	References
Japan		4.6	1958	800	6–14	Shiokawa Y., 1975
		0.1	1971	800	6–14	
New Zealand (east coast) Maori population	535		1963–1969		10–14	Stanhope J.M., 1976
Other population	55				10–14	
Hong Kong	23.1		1972	4,000	all ages	Wu R.W.Y., 1975
Taiwan		1.4	1970–1971	5	1–19	Lue H.Ch., 1975
Indonesia		1.0	1970	4	1–14	Hanafiah A., 1973
Singapore	92		1971	2,000	all ages	Loh T.F., 1975
India Different localities		1.4 to 15.0	1972–1974	28	1–14	Padmavati S., 1974
North		1.65		33	all ages	Berry J.N., 1972
South		6.48	1974	6.3	1–14	Koshi G., 1975
Iran Teheran	58.51		1971, 1973	56		Gharagozloo R., 1975
Abadan	100		1971	70	all ages	Partow A., 1972
Cyprus	27 to 43		1962–1972	150	all ages	Kalbian V.V., 1972
Egypt		10	1966–1972		6–12	El Kholy A., 1972
Algeria		15	1970	470	6–14	Mostefai M.C., 1972

c) Each attack of rheumatic fever, whether primary or recurrent, is preceded by symptomatic or asymptomatic group A streptococcal infection.

d) The efficacy of a secondary streptococcal disease prevention programme in rheumatic individuals or groups of them can best be monitored through periodic checking by means of microbiological techniques.

e) Primary prevention of rheumatic fever and acute glomerulonephritis on a basic (peripheral) health-care level requires clinical work in conjunction with laboratory services if a high standard of health and economic efficacy is to be achieved.

f) Microbiological examination for streptococcal infection is relatively simple and inexpensive and can be performed on a wide scale. However, it requires constant control to ensure a standard level of work guaranteeing reliability of laboratory results.

g) The microbiological laboratory component is an essential part of any streptococcus control programme. The payback when cost versus benefit is assessed is substantial.

The *World Health Organization* has been directing its interest to the streptococcus problem for over 10 years. The activities of WHO in rheumatic fever and rheumatic heart disease have been stimulated by the data reported on these diseases from countries in hot climatic zones. Requests for assistance in national prevention programmes have been expressed and suggestions for corresponding projects coordinated by the WHO advanced at several meetings.

Strengthening of laboratory services is one of the WHO activities aimed at an improvement and further advancement of laboratory facilities. In general, this programme, which also serves for the provision of better information on the epidemiology and control of these diseases, consists of two major components: the introduction of standard laboratory procedures in laboratories already existing and the establishment of adequate laboratory facilities in as many districts of a country or geographical region as necessary. Emphasis is laid on the simplicity and reliability of methods and on as much independence as possible from services outside the laboratory.

The *strategy* for establishing such facilities, in particular in developing countries, and of integrating them into existing health-care services, might be as follows:

The realistic approach would be to build up a *general system* [14] consisting of three levels:

a) At the *national level, peripheral* services within the country would consist of a network of points for collecting and transporting specimens located at health-care centres and of a network of laboratories having the basic microbiological methods for the diagnosis of streptococcal infections, rheumatic fever and acute glomerulonephritis at their command.

b) At the national level, also, a *central* national streptococcus laboratory would have charge of specialized laboratory procedures, carry out research and provide assistance to the peripheral level in terms of expert guidance, reference work and training.

c) At the *WHO level*, WHO collaborating centre(s) would assist with WHO projects coordinated by headquarters and regional offices, would cater for standardization of methods, pursue research programmes of their own and work in close cooperation with national streptococcus centres, assisting them with reference work and training. The WHO collaborating centres would cooperate closely with other advanced streptococcus research laboratories (Table V).

Which *methods are the most appropriate* and convenient as identification and classification techniques in streptococcal bacteriology and serology?

The decision as to the use of a *particular laboratory procedure* depends on the purpose of the examination and on the technical facilities of the laboratory. Because the diagnosis of streptococcal infections is already mostly carried out in institutions belonging to the primary health-care structure, it is desirable to give preference to *simple methods*. These should be fully applicable but practicable in laboratories not possessing advanced technical facilities. District laboratories or

Table V. Strategic plan for laboratory services in the streptococcus problem.

Activities-Functions-Links		
National level		**WHO level**
Peripheral	Central	
Collection and transport of specimens	Specialized laboratory procedures	Assistance in HQ and regional WHO projects
Basic procedures of microbiological diagnosis	Research	Standardization of methods
	Assistance to peripheral level: expert guidance, reference work, training	Research
		Assistance to national central level: reference work, training

national laboratories should be versed in, and employ, *specialized methods*. The system of quality control of laboratory work is essential. The professional level of work should be subject to upgrading as new contributions in the field arise.

Several facets of *laboratory technicalities* will be presented and discussed at this meeting. I wish to make a few comments on particular points.

Concerning the *transport of material* to the laboratory for microbiological examination, a satisfactory state of the specimen is a prerequisite for a reliable result. Various materials for holding specimens and several ways of preserving bacteria during transport have been advocated. Although the deleterious effects of humidity and higher temperature have been conclusively established, more information is still needed on their real role under natural conditions. Only after this information has been provided can an optimal system widely applicable for use in different climatic conditions and over considerable distances be proposed. This is essential for overcoming the risks relating to delay in inoculating specimens. Field trials rather than laboratory models simulating natural conditions are required for this purpose. Transport is a particular problem in hot climatic areas.

The *cultivation and isolation* of haemolytic streptococci does not, as a rule, present any technical problem if proper media are available and the material has been adequately treated. In peripheral laboratories, however, more attention should be paid to species of *viridans* and nonhaemolytic streptococci. Anaerobic streptococci, the peptostreptococci, could do with provision of more information on their biological properties and role in human pathology. The findings from experimental studies on L forms collected over the last few years have turned out to be less applicable in diagnostic practice than expected. Whether this is due to an inadequacy in the methods employed for L form search in human pathology, to a lack of a pathogenic role of L forms or to their absence from human and animal diseases remains to be shown.

Classification of streptococci into *species* has largely been established. A number of methods and modifications have been developed for serological grouping according to the Lancefield scheme. The technological requirement for extraction of the group antigen from the cells is an obstacle to wide routine use of serological grouping at the peripheral laboratory level. Moreover, some procedures require potent sera. For example, a grouping serum precipitating poorly with HCl extract in the capillary test may be found satisfactory in the ring test with formamide extract. Among the inflated number of grouping methods a real breakthrough is represented by the co-agglutination method. However, the identification of particular groups, for example group D, by this method requires still more work, although this is the case with many other grouping techniques as well. The status of some groups or group candidates still needs to be verified. It must be shown conclusively that these are not antigenically related to or identical with already existing groups, e.g., group D.

Concerning the nongroupable β, α and nonhaemolytic streptococci, much progress has been made in recent years in their identification and classification [15, 16]. This is true, for example, of *Streptococcus milleri*, *S. mutans*, *S. mitior* and *S. sanguis*. Their pathogenic role is obvious, although, strictly speaking, a simple isolation may not justify the microbiologist in ascribing an aetiological role to a microbe. To meet Koch's postulates fully, we need more knowledge on the biological properties of these species, as well as on antibodies in diseases caused by them. In particular, new and more specific tests would be very helpful for allocating to classes those strains in which classification problems are encountered when current biochemical tests are used.

Typing of group A streptococcus as a major human pathogen has largely been elaborated in the past and is subject to a renaissance of interest at present [17].

Three typing systems are now available (Table VI). M typing with good unabsorbed sera (or absorbed sera if required) in a double diffusion test is a simplification of the typing method. It facilitates typing of large numbers of strains. SOF typing considerably increases the typability of group A strains. However, the gap between advanced streptococcal laboratories and district or central (national) laboratories that practice typing seems to be widening. This is due not only to the increasing sophistication of typing techniques but also to the appearance of new types.

The use of the typing method depends on the facilities of the laboratories. It would be helpful to create better provisions for typing serum production, at least of the prevalent types. This would become urgent once a streptococcus vaccine had been developed and made available for human use.

Various laboratory tests have been devised for the *titration of streptococcal antibodies*. Some of the tests can be standardized to a degree ensuring comparability of results obtained in different laboratories (e.g., the ASO test). Other tests may give valuable information in a particular laboratory but are of rather limited value as regards comparability. An essential point is the priority of some tests over others with regard to their diagnostic value.

Although the *antistreptolysin O test* has been standardized and various modifications of the techniques involved elaborated, the technicalities of the test still produce some problems which require continuous attention and frequent quality control in

Table VI. Typing systems for group A streptococci.

M typing	T typing	SOF typing
	IDENTIFY	
M type	T pattern	SOF type
	DIFFERENTIATE BETWEEN	
M types 1 to 65	T patterns (include one or	SOF types
(and candidates for new M types)	more M types)	(identical with M types)
Numbers not used:	1 1	2, 4, 9, 11, 13, 22, 25, 28,
7, 10, 16, 20, 21, 35, 64	3/13/B$_{3264}$ . . 3, 13, 33, 39, 41,	48, 49, 58, 59, 60, 61, 62, 63
	43, 52, 53, 56	
Partial antigenic relationship	4/28 . . . . 4, 24, 26, 28, 29,	
	46, 48, 60, 63	
in types	5/12/27 . . . . . 5, 11, 12, 27, 44,	
	61, 62	
13 + 48	6 6	
2 + 48	8/25/Imp. 19 . 2, 8, 25, 31, 55,	
	57, 58, 59, 65	
3 + 12	9 9	
33 + 41 + 43 + 52	14/49 . . . . 14, 49	
	15/23/47 . . . . 15, 17, 19, 23,	
	30, 47, 54	
	18 18	
	22 22	

order to maintain reliability. There is an urgent need for local standards of anti-streptolysin O wherever streptolysin is produced. Such standards must be precisely related to the international standard.

The spectrophotometric method based on evaluation of 50 % haemolysis is preferable but the visual method, if an appropriate serum dilution range is used, is fully adequate too.

To assess the *antideoxyribonuclease B* titre in human serum a micromethod is generally used. It employs the deoxyribonucleic acid-methyl green complex as the substrate and identifies the end point by colour change.

Because of the lack of an international deoxyribonuclease B standard, the numerical values of serum titres have local significance only and cannot be compared with titres from elsewhere if different reagents have been used. However, collaborative work under WHO sponsorship has been started and an international standard of antideoxyribonuclease B is likely to be available in the near future.

Other tests are, for example, assays for antihyaluronidase, antidiphosphopyridine-dinucleotidase and antistreptokinase. The recently advocated Streptozyme test, though technically simple, requires more detailed evaluation before a conclusion concerning its value and usefulness can be drawn.

Streptococcal infections and their sequelae have been confirmed as being of *public health importance* in a number of countries. The term 'importance' has to be considered in relation to other health problems in a particular population, since priorities have to be determined. The availability of laboratory facilities for setting up reliable diagnosis and installing efficient therapy and prevention programmes fully justify the employment of microbiological methods in connection with the streptococcus problem on a larger scale than at present.

REFERENCES

1. Duben, J., Beránek, M., Kubcová, M., Jelínková, J., Vojtěchovská, H. and Rotta, J. (1973): Morbidity from streptococcal infections in a general population (a five-year study). *J. Hyg. Epidemiol. Microbiol. Immunol.*, *17*, 43.
2. El Kholy, A., Sorour, A.H., Houser, H.B., Wannamaker, L.W., Robins, M., Poitras, J.-M. and Krause, R. (1973): A three-year prospective study of streptococcal infections in a population of rural Egyptian school children. *J. Med. Microbiol.*, *6*, 101.
3. Siew Tin Chen, Dugdale, N.E. and Pathucheary, S.D. (1972): Beta-haemolytic streptococcal carriers among normal schoolchildren. *Trop. Geogr. Med.*, *24*, 257.
4. Baldovin-Agapi, C.: Porteurs de streptocoques hémolytiques et immunité antistreptococcique dans le Vietnam du Nord. *Arch. Roum. Pathol. Exp. Microbiol.*, *21*, 719.
5. Duben, J., Jelínková, J., Jelínek, J. and Rotta, J. (1979): A prospective study on streptococcal pharyngitis among a town population. *J. Hyg. Epidemiol. Microbiol. Immunol.*, in press.
6. Duben, J., Jelínková, J., Míčkova, S., Havlíčkova, H., Vojtěchovská, H., Beránek, M. and Rotta, J. (1978): Nine-year study of streptococcal infections in a sample of the general population. *J. Hyg. Epidemiol. Microbiol. Immunol.*, *22*, 162.
7. Meyers, R.M. and Koshi, G. (1963): Beta-haemolytic streptococci in a survey of throat cultures in an Indian population. *J. Publ. Health*, *51*, 1872.
8. Ogunbi, O. (1971): A study of beta-haemolytic streptococci in throats, noses and skin lesions in a Nigerian (Lagos) urban population. *J. Nigeria Med. Assoc.*, *1*, 59.
9. Prakash, K. *et al.* (1967): Increasing incidence of group B beta-haemolytic streptococci from human sources. *Ind. J. Med. Res.*, *6*, 506.
10. Koshi, G., Mammen, A., Feldman, D.B., Bhakthaviziam, C. and Myers, R.M. (1976): A preliminary report on beta-haemolytic streptococci and antistreptolysin O (ASO) titres in pyogenic skin infections in children, with a case report of acute glomerulonephritis following repeated skin infections. *Ind. J. Med. Res.*, *9*, 920.
11. Valkenburg, H.A. (1971): Streptococcal pharyngitis in the general population. II. The attack rate of rheumatic fever and acute glomerulonephritis in patients not treated with penicillin. *J. Infect. Dis.*, *124*, 348.
12. Kopecká, B., Šrámek, J., Bošmanský, K., Ringel, J. and Peychl, L. (1979): Rheumatic heart disease: situation in Czechoslovakia with special reference to long term prognosis for rheumatic carditis. In press.
13. Strasser, T. and Rotta, J. (1976): The control of rheumatic fever and rheumatic heart disease: an outline of WHO activities. *WHO Chron.*, *27*, 49.
14. WHO Memorandum (1978): Recent advances in rheumatic fever control and future prospects. *Bull. W.H.O.*, *56*, 887.
15. Parker, M.T. and Ball, L.C. (1976): Streptococci and aerococci associated with systemic infection in man. *J. Med. Microbiol.*, *9*, 275.
16. Facklam, R.R. (1974): Characteristics of *Streptococcus mutans* isolated from human dental plaque and blood. *Int. J. Systematic Bacteriol.*, *24*, 313.
17. Rotta, J. (1972): Prospects for improved approaches to and reagents for identification of streptococci. In: *Streptococci and Streptococcal Diseases*, Chapter 16, p. 267. Editors: L.W. Wannamaker and J.M. Matsen. Academic Press. New York.

Pathogenicity of streptococci

E. L. Randall
Evanston Hospital, Evanston, U.S.A.

Prior to the antibiotic era, serious life-threatening infections were most commonly produced by streptococci, staphylococci, pneumococci and meningococci. After the introduction of antimicrobial agents, these organisms were more readily controlled. Staphylococci appeared as a serious cause of nosocomial disease in the late 1950s and the early 1960s, after which Gram-negative rods emerged as the most common agents of serious infections, especially those of a nosocomial nature.

The causative agents isolated from clinically significant cases of bacteraemia were examined from patients at two separate institutions where the author was a clinical microbiologist, Thomas Jefferson University Hospital (TJUH), Philadelphia, (1955 to 1971) and Evanston Hospital (EH), Evanston, (1974 to 1978).

As can be observed in Table I, Gram-negative organisms accounted for 44.5 %, Gram-positive organisms for 46.4 %, anaerobes for 6 % and fungi for 3.1 % of the 4,394 bacteraemia cases at TJUH. Streptococci were isolated from approximately 20 % of the cases. At EH, Gram-negative organisms were isolated from 65.7 %,

Table I. Bacteraemia – total organisms.

	TJUH (1955–1971)		EH (1974–1976)	
	Number	% of total	Number	% of total
Gram-negative organisms				
Enterobacteriaceae	1,488	33.8	581	48.8
Non-fermenters	386	8.8	66	5.5
Small rods	60	1.4	27	2.3
Neisseria	21	0.5	1	0.1
Total	1,955	44.5	675	56.7
Gram-positive organisms				
Staphylococci	1,108	25.2	204	17.1
Streptococci	883	20.1	184	15.5
Aerobic rods	48	1.1	3	0.3
Total	2,039	46.4	391	32.9
Miscellaneous				
Anaerobes	262	6.0	69	5.8
Fungi	138	3.1	55	4.6
Total	400	9.1	124	10.4
Total	4,394	100.0	1,190	100.0

Gram-positive from 32.9 %, anaerobes from 5.8 % and fungi from 4.6 % of cases. Streptococci were isolated from 15.5 % of bacteraemias.

Staphylococcus aureus was the most important cause of bacteraemia in the late 1950s and early 1960s at TJUH but steadily declined in the middle and late 1960s. In 1956, 55 % of positive blood cultures contained *S. aureus*, in 1963, 30 %, in 1965, 20 %, and from 1966 to 1971 approximately 8 %. Collectively, Gram-negative rods were isolated from 24 % of positive blood cultures in 1956. They steadily increased to a high of 56 % in 1957 and levelled off in 1970 and 1971 to 48 %. During these years the percentage of streptococci from bacteraemia cases fluctuated slightly, varying from 10 to 11 % in certain years to as high as 24 % in another.

In Table II, groups of streptococci isolated from blood cultures from the two institutions in the years indicated are depicted. The pneumococcus was the most common species isolated from both institutions. The second most common at TJUH were viridans streptococci but at EH it was group D streptococci. A possible explanation for this might be the more frequent occurrence in the last few years of group D streptococci as causative agents of subacute endocarditis than in earlier years. The majority of organisms listed as not group A from both hospitals were probably group B streptococci, as most were isolates from neonatal populations. Over the years from 1955 to 1978 the percentage of group A streptococci has remained fairly constant at both hospitals while the percentage of isolates of group B streptococci has been slowly increasing.

Table II. Bacteraemia – streptococci.

| | TJUH (1955–1972) | | EH (1974–1978) | |
	Number	% of total	Number	% of total
Group A	57	10.2	15	8.4
Not group A	66	11.8	14	7.9
Group B	15	2.7	20	11.2
Group D	54	9.7	48	27.0
Group G	5	0.9	2	1.1
Viridans	153	27.5	28	15.7
Pneumococci	207	37.2	51	28.7
Total	557	100.0	178	100.0

The causative agents of neonatal bacteraemia at the two institutions are cited in Table III. Enterobacteriaceae were most common, followed by streptococci and staphylococci. A breakdown of the various types of streptococci isolated can be seen in Table IV. At TJUH, 47.4 % of the streptococcal isolates were viridans streptococci. It could not be determined how many of these were true isolates. It was often believed that most were contaminant. At both institutions, the non-group A isolates were most likely group B, which would indicate that 38.1 % of the streptococcal isolates at TJUH and 60.7 % at EH were *Streptococcus agalactiae*.

To indicate the differences observed in various age-groups the organisms isolated from the paediatric population are represented in Table V. By far the most common

Table III. Neonatal bacteraemia – total organisms.

	TJUH (1955–1972)		EH (1974–1978)	
	Number	% of total	Number	% of total
Gram-negative organisms				
Enterobacteriaceae	128	38.0	44	43.1
Non-fermenters	29	8.6	9	8.8
Haemophilus	9	2.7	0	0.0
Total	166	49.3	53	51.9
Gram-positive organisms				
Staphylococci	68	20.2	13	12.8
Streptococci	85	25.2	29	28.4
Total	153	45.4	42	41.2
Miscellaneous				
Anaerobes	13	3.8	2	2.0
Fungi	5	1.5	5	4.9
Total	18	5.3	7	6.9
Total	337	100.0	102	100.0

isolate in EH was *Haemophilus influenzae* which comprised 36 of the 37 small Gram-negative rods recovered. At TJUH the two most prevalent organisms were staphylococci and streptococci.

An analysis of streptococci (Table VI) reveals the pneumococcus to be the pathogen most frequently isolated from blood cultures in the paediatric age-group. As would be expected, group B streptococci are not commonly isolated as pathogens from this particular population.

Table IV. Neonatal bacteraemia – streptococci.

	TJUH (1955–1972)		EH (1974–1978)	
	Number	% of total	Number	% of total
Group A	0	0.0	3	10.7
Not group A	21	27.6	3	10.7
Group B	8	10.5	14	50.0
Group D	9	11.9	4	14.3
Viridans	36	47.4	4	14.3
Pneumococci	2	2.6	0	0.0
Total	76	100.0	28	100.0

Table V. Paediatric bacteraemia – total organisms.

	TJUH (1955–1972)		EH (1974–1978)	
	Number	% of total	Number	% of total
Gram-negative organisms				
Enterobacteriaceae	55	19.2	7	11.7
Non-fermenters	17	6.0	1	1.65
Small rods	21	7.4	37	61.7
Neisseria	9	3.1	1	1.65
Total	102	35.7	46	76.7
Gram-positive organisms				
Staphylococci	97	33.9	6	10.0
Streptococci	82	28.7	8	13.3
Total	179	62.6	14	23.3
Miscellaneous				
Anaerobes	4	1.4	0	0
Fungi	1	0.3	0	0
Total	286	100.0	60	100.0

From 1967 to 1972, organisms isolated from various body fluids at TJUH were examined. These fluids are listed in Table VII. The numbers of each fluid examined and the percentage from which Gram-positive cocci were isolated and, in addition, the percentage of total Gram-positive cocci that each species represents are shown. Staphylococci were most prevalent, followed by viridans and group D streptococci. Of the β-haemolytic streptococci, groups other than A were more common than group A:

In recent years more importance is being attached to the taxonomy of α- and non-haemolytic streptococci belonging to the viridans group. As more exact iden-

Table VI. Paediatric bacteraemia – streptococci.

	TJUH (1955–1972)		EH (1974–1978)	
	Number	% of total	Number	% of total
Group A	10	12.2	0	0
Not group A	1	1.2	0	0
Group D	3	3.7	1	12.5
Viridans	16	19.5	1	12.5
Pneumococci	52	63.4	6	75.0
Total	82	100.0	8	100.0

Table VII. Body fluids.

			Percentage of total Gram-positive cocci					
			TJUH (1967–1972)					
Body fluid	% of isolates	% with Gram-positive cocci	Staphylo-cocci	Group A strept.	Not A strept.	Viridans strept.	Pneumo-cocci	Group D strept.
Dialysis	353	34	45	1.5	1.5	31	1	20
Peritoneal	758	35	34	2	8	30	1	25
Pleural	385	49	49	2	3	20	9	17
Spinal	218	42	42	4	9	7	14	24
Synovial	85	73	66	6	2	13	0	13
Amniotic	158	45	38	3	11	41	0	7

tification procedures are utilized for this group the more important species isolated from pathogenic conditions may be better appreciated.

As rapid methods are now available for the routine grouping of β-haemolytic streptococci the significance of groups other than A and B in pathogenic conditions will, it is hoped, be better understood.

Discussion

J. A. Morello (Chicago, U.S.A.): There was an extremely high incidence of viridans streptococci in the fluids. Was there a significant infection from these?

E. L. Randall: At one time when we went over these data this was, of course, very disturbing. It is difficult even now really to appreciate the importance of viridans streptococci especially from peritoneal and various other body fluids. As I mentioned, most of these were mixed infections where perhaps anaerobes or Enterobacteriaceae played a more important role, and where these viridans streptococci were just opportunist. Indeed, on many occasions I do not think that they were the significant cause of the infection, but were simply organisms that tagged along.

R. R. Facklam (Atlanta, U.S.A.): The more recent developments concerning identification of the species are probably some of the most important things which will help to answer this question of the significance of viridans streptococci in clinical material. One thing which has evolved from the schemes devised by myself and by Parker and Ball at Colindale is that there may be specific viridans in various diseases, such as *Streptococcus milleri*, in the abscesses and normally sterile body fluids similar to those discussed by Dr Randall, but until more data are gathered on the relationship of the various viridans species in these kinds of diseases, the answer will not be known.

Phenotypic aspects of resistance to macrolide and related antibiotics in β-haemolytic group A, B, C and G streptococci

T. Horodniceanu, L. Bougueleret and F. Delbos
Streptococcus Reference Centre, (Medical Bacteriology Unit), Institut Pasteur, Paris, France

Streptococcal infections still represent a public health problem although considerable improvements have been made in their diagnosis, therapy and prevention. Lancefield group A, B, C and G β-haemolytic streptococci are the most common pathogens in man, primarily in infections of the upper respiratory tract and the skin and in neonates. They are usually very susceptible to penicillins, macrolide antibiotics and related drugs, tetracyclines, chloramphenicol and vancomycin.

The appearance of drug-resistant strains in Lancefield group A, B, C and G streptococci has been previously reported, especially for macrolide, lincosamide and streptogramin B (MLS resistance), and for tetracycline and chloramphenicol antibiotics [1–4].

Evidence for the presence of R-plasmids in group A and B strains has been previously reported [5–7]. Conjugative transfer of R-plasmids harboured by strains of groups A and B has recently been reported [8–10]. Evidence of the existence of conjugative R-plasmids in groups C and G has been demonstrated in our laboratory (report in preparation).

In this study, we report the different phenotypic aspects of MLS resistance in group A, B, C and G streptococci. The purpose of our study was, firstly, to provide information on MLS resistance occurring in strains already resistant to erythromycin or lincomycin and, secondly, to emphasize that, in such cases, no macrolide or related drug, such as lincosamides and streptogramin B, should be used to treat streptococcal infections.

MATERIALS AND METHODS

Bacterial strains

A total of 30 clinical specimens of β-haemolytic streptococci were isolated in different laboratories in France from nasopharyngeal, skin and vaginal swabs and from pus, blood and cerebrospinal fluid. The streptococci were identified by Fuller's serological grouping test [11]. They belonged to group A (six strains), group B (13

Table I. Summary of the phenotypic features of β-haemolytic streptococcal resistant strains.

		Category		
I	II	III c	IV d	V
Mac ^{R}	Mac ^{R}	Mac ^{R}	Mac ^{I}	Mac ^{I}
Lin ^{R}	Lin ^{R} b	Lin ^{R} b	Lin ^{S} e f	Lin ^{R}
SgB ^{R}	SgB ^{R}	SgB ^{I}	SgB ^{I}	SgB ^{I}
A457	B96 g h	A451	G45	A456
G43	B97 g h	A453 i	G47	B101
G49 g h	B98 g h	A454 g	G48	B111
G50	B109 g	A455 g h		B114
	B110 g h	B116 e		G42 g
	B113 g h	B121 e g		G44 g
	B115 g h	B122 g h		
	C87 g h	C88 e		
	G41 g h			

Abbreviations: Mac, macrolides; Lin, lincosamides; SgB, streptogramin B; R, resistant; I, intermediate; A, serogroup A; B, serogroup B; C, serogroup C; G, serogroup G
b Diminution of the growth inside the inhibition zone of lincomycin
c Interaction between Mac and Lin
d Interaction between Mac and Lin and between Mac themselves
e Susceptible to lincomycin after 18 hours, resistant after 36 hours of incubation
f Zone-like phenomenon for lincomycin
g Conjugative transfer of MLS markers into streptococcal recipients
h Physical evidence of plasmid DNA
i Susceptible to Lin even after 36 hours of incubation, with marked interaction between erythromycin and Lin

strains) group C (two strains) and group G (nine strains) (Table I). All strains were antibiotic-resistant and had been referred specially to us, primarily from Strasbourg, Paris, Toulouse and Nantes.

Media

The media used were brain-heart infusion broth (BHI; Difco), brain-heart infusion agar (BHIA; Difco) and Mueller-Hinton agar (MH; Pasteur Institute) supplemented with 5 % horse serum or 5 % horse blood.

Antibiotic susceptibility tests

Minimum inhibitory concentrations (MICs) were determined as previously described [6] for erythromycin (Abbott), lincomycin (Upjohn), chloramphenicol (Roussel) and tetracycline (Rhône-Poulenc).

Disc diffusion zone determinations were performed using overnight broth cultures diluted 1 : 10 for serogroups A, C and G and 1 : 200 for group B strains and spread on plates. For standard determinations (distance between discs was 30 mm) MH with 5 % horse blood was used. Discs were as follows: penicillin G (Pc), chloramphenicol (Cm), tetracycline (Tc), erythromycin (Em), streptogramin B (SgB), rifampicin (Rif) and fusidic acid (Fus). In addition, all drug-resistant strains were tested in BHIA with 5 % horse serum. In such cases the distance between discs was reduced to 20 mm. The antibiotics used and their disposition on each plate correspond to the pattern shown in Figure 1. Abbreviations of antibiotics are given in Table II. Plates were incubated aerobically at 37 °C for 18 and 36 hours.

Table II. Levels of antibiotic resistance of group A, B, C and G streptococci tested by MIC and susceptibility discs.

Antibiotics	MIC (μg/ml)	Susceptibility discs b	Interpretation c
	0.015–0.03	28–30	S
Em	0.5–64	15–22	I
	125–2,000	6–12	R
		25–30	S
Sp	NT	15–22	I
		6–12	R
Om	NT	23–> 25	S
		6–12	R
	0.06–0.12	28–30	S
Lm	0.5–2	15–25	I
	125–500	6–12	R
		24–28	S
Cl	NT	14–22	I
		6–12	R
		20–27	S
SgB	NT	13–18	I
		6–12	R

Abbreviations: Em, erythromycin; Sp, spiramycin; Om, oleandomycin; Lm, lincomycin; Cl, clindamycin; SgB, streptogramin B
b Diameters in mm of inhibition zone (6 mm, no inhibition zone)
c S, susceptible; R, resistant; I, intermediate
NT, not tested

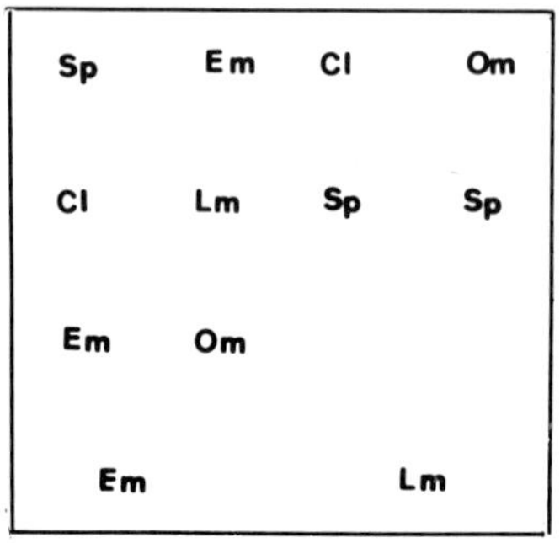

Fig. 1. Antibiotics used and their disposition on each plate (distance between discs: 20 mm). For abbreviations see Table II.

124

RESULTS

Antibiotic resistance tested by MIC

The levels of susceptibility and resistance measured by MIC for Em and Lm are shown in Table II. Some strains appeared to have intermediate figures for these drugs. Intermediate susceptibility to Em and Cl has been previously reported in *Streptococcus pyogenes* (group A) [12] and recently in *Streptococcus pneumoniae* resistant strains [13].

Antibiotic resistance tested by susceptibility discs

The figures given in Table II defined fully susceptible, intermediate and resistant strains for macrolides, lincosamides and SgB as tested by the standard method. In several cases, when Em and Lm discs were placed side by side the circular zone diameter of Lm was reduced by Em, producing an oval zone for Lm (Fig. 2). Moreover, in many resistant strains a visible borderline between Em and Lm was observed instead of a homogeneous bacterial growth (Fig. 2). These borderlines were straight, following the antibiotic diffusion front. In most cases deformation of the inhibition zone and borderlines was more easily seen when the distance between discs was reduced to at most 20 mm; borderlines were visible only on transparent media (supplemented with horse serum). In order to define the mechanism involved in this interaction macrolides and lincosamides were combined so as to obtain close contact between the diffusion zones of two drugs. The results are shown in Figures 3 and 4. An antagonism phenomenon was observed between macrolides and lincosamides (Fig. 3) and between Em or Om and Sp (Fig. 4). Usually interactions between macrolides and Cl were less evident than between macrolides and Lm.

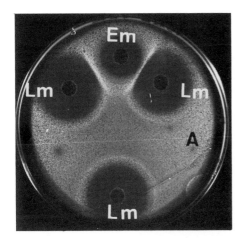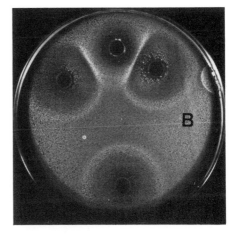

Fig. 2. Strain B116; deformation of the circular inhibition zone of lincomycin (Lm) by erythromycin (Em) and borderline between the two drugs. A: 18 hours of incubation. B: 36 hours of incubation.

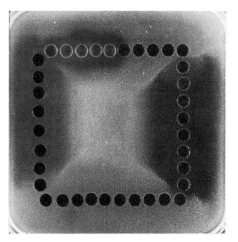

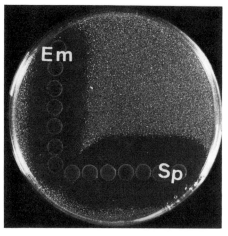

Fig. 3. Strain A455: antagonism between erythromycin and clindamycin (bottom left); spiramycin and lincomycin (bottom right); streptogramin B and clindamycin (top left); and between oleandomycin and lincomycin (top right).

Fig. 4. Strain G47: antagonism between erythromycin (Em) and spiramycin (Sp).

Phenotypic aspects of MLS resistance

The 30 resistant strains were divided into different classes on the basis of the inhibition zones obtained for macrolides and lincosamides and of the interactions (borderlines) between these drugs. This yielded the following five categories.

I – In four cases (one group A and three group G) the growth of the strain was homogeneous and no inhibition zone was observed around the susceptibility discs: growth was in contact with both macrolides and lincosamides (Fig. 5); a very small inhibition zone (diameter 8–10 mm) was observed for SgB.

II – In nine cases (seven group B, one group C and one group G) the same results

Fig. 5. Strain A457: example of category I (see text).

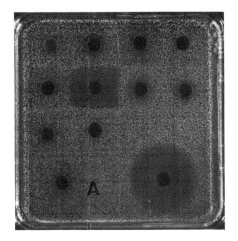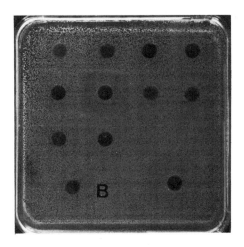

Fig. 6. Strain B96: example of category II (see text). A: 12 hours of incubation. B: 18 hours of incubation.

as in the first class were obtained but a diminution of growth was visible inside the inhibition zone of Lm (Fig. 6). This diminution was more visible after 10–12 hours of incubation. For SgB a very small inhibition zone was observed (diameter 10–12 mm). In addition, as shown in Figure 3, an antagonism between macrolides and Lm was noted.

III – In eight cases (four group A, three group B and one group C), although growth was in contact with macrolide discs and only diminished inside lincosamides, it did not appear to be homogeneous because of the resulting interactions, especially between macrolides and lincosamides (Fig. 7). In three cases (groups B and C), as shown in Figure 2, resistance to Lm appeared only after 36 hours of incubation, giving a false impression of susceptibility after 18 hours. For SgB an intermediate resistance was observed.

IV – In three cases (group G) the resistance to Em and Sp was intermediate and

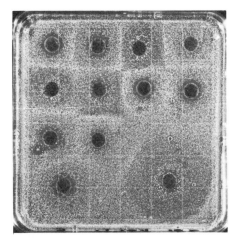

Fig. 7. Strain A451: example of category III (see text).

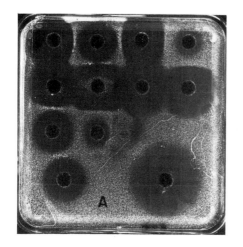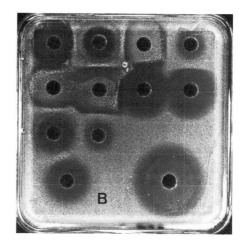

Fig. 8. Strain G48: example of category IV (see text). A: 18 hours of incubation. B: 36 hours of incubation (note also the zone-like phenomenon of resistance to lincomycin (Lm) and clindamycin (Cl).

the strains seemed to be susceptible to lincosamides after 18 hours of incubation; interactions between all drugs were visible (Fig. 8). Again, these strains appeared to be resistant to Lm and Cl only after 36 hours of incubation and, in addition, a zone-like phenomenon was observed (Fig. 8B). The zonal pattern of resistance to Lm was previously reported for resistant group A streptococci [2]. For SgB an intermediate resistance was observed.

V – In six cases (one group A, three group B and two group G) resistance to Em and Sp was intermediate and no inhibition zone was observed for lincosamides (Fig. 9). Growth was homogeneous, without any change after 36 hours; antagonism between Em and Sp could be observed only when discs were placed as shown in Figures 3 and 4. For SgB an intermediate resistance was observed.

In the last two categories no inhibition zone was observed for Om. In preliminary

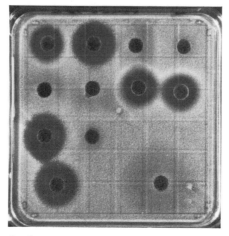

Fig. 9. Strain A456: example of category V (see text).

studies of the resistant strains presented here antagonism between SgB and lincosamides was observed only in a few cases and antagonism between Em and SgB was noted only in one strain (G44). A summary of the most important phenotypic features of the 30 resistant strains is given in Table I.

DISCUSSION

Resistance to Em in clinical isolates of *Staphylococcus aureus* is generally associated with a chemical alteration of ribosomal structure which specifies coresistance to macrolides, lincosamides and SgB. The biochemical basis for MLS resistance is generally associated with a specific N^6-dimethylation of adenine in 23 S ribosomal ribonucleic acid [14]. Resistance to Em in streptococci involves the same MLS-type resistance with different phenotypic aspects. Weisblum and co-workers recently reported the possibility of a common origin for Em-resistance determinants in unrelated strains of pathogenic staphylococci and streptococci for which exchange of genetic material has not been demonstrated [15].

In France, the emergence of multiresistant β-haemolytic streptococci has only recently been reported [6, 16] in comparison with other countries [1, 17, 18]. In spite of the relatively frequent use of macrolides and related drugs, MLS resistance occurs infrequently (0.5–2 % in group A, B C and G streptococci). In contrast, resistance to macrolides and lincosamides of *S. pyogenes* (group A) is very high in Japan (more than 60 %) [19].

The two major types of Em resistance, constitutive and inducible [20], were found among the five strains studied (data not shown): two strains of group B (B96 and B114) were constitutive and a group A strain (A451) and two group B strains (B109 and B115) were inducible. It is too early to establish a relation between the category and the type of resistance. Further studies involving all resistant strains would allow it to be determined if such relationships really exist.

Genetic and molecular studies of the 30 antibiotic-resistant streptococci of groups A, C, G (in preparation) and B [10] were undertaken after the transfer of resistance markers into streptococcal recipients. Conjugative transfer of MLS-resistance markers was obtained from 16 wild strains into group B and/or group D recipients. Plasmid DNA could be demonstrated in 11 cases, eight of which belonged to category II (Table I). Molecular weights of plasmids ranged from 17 to 25 megadaltons depending on the absence or presence of Cm and/or Tc markers [10 and report in preparation].

The possible relation of MLS streptococcal R-plasmids was studied by DNA-DNA hybridization [7, 21]. El-Solh and co-workers [7] obtained more than 90 % nucleotide sequence homology between pIP501 [6] and pIP612 R-plasmids harboured by two different group B strains isolated in 1975 in Strasbourg and an MLS plasmid, pIP613, carried by a *Streptococcus faecalis* isolated in 1963 in Paris [22]. Moreover, a high degree of homology was also obtained with an MLS plasmid, pIP646, harboured by a group C strain isolated in 1976 in Strasbourg and with two other MLS plasmids, one harboured by a group A strain [5] isolated in Canada and the second by a group D strain [23] isolated in the United States (N. El-Solh, personal communication). These results support the hypothesis that very similar if not identical plasmids might be disseminated among different species of streptococci [7].

129

A good knowledge of MLS-type resistance is very important for the correct treatment of streptococcal infections. From data presented here it is evident that in six cases at least, a false susceptibility might have been diagnosed if incubation had not been prolonged to 36 hours. These results suggest that resistance to Lm and Cl, at least in these strains, appears to be auto-inducible (Lm discs were placed far enough from the influence of other drugs to rule out interactions between macrolides and lincosamides). Consequently, in such situations, as well as in cases of intermediate resistance to macrolides and SgB, cross-resistance between all these drugs should be considered. Treatment with such antibiotics is, therefore, inadvisable. Penicillin is usually the antibiotic of choice in the treatment of β-haemolytic group A, B, C and G streptococcal diseases. In penicillin-sensitive patients for whom penicillin is contra-indicated Em or Lm is, generally, a satisfactory alternative. The emergence of strains resistant to Em and Lm, the prevalence of Tc-resistant strains and the appearance of high-level resistance to aminoglycoside antibiotics in β-haemolytic streptococci (report in preparation) may make it difficult, in some cases, to provide appropriate therapy.

SUMMARY

Thirty antibiotic-resistant clinical isolates of β-haemolytic streptococci (groups A, B, C and G) were investigated for their phenotypic aspects of resistance to macrolide (erythromycin, oleandomycin, spiramycin), lincosamide (lincomycin, clindamycin) and streptogramin B antibiotics (MLS resistance). Five different categories were found with regard to the inhibition zone around susceptibility discs and/or the existence of interactions between these drugs. The interaction was found to be an antagonism phenomenon. In all strains a coresistance to MLS drugs was observed. In six strains, resistance to lincosamides was evident only after 36 hours of incubation.

ACKNOWLEDGEMENTS

This work was supported by grant 78.4.137.1 from the Institut National de la Santé et de la Recherche Médicale (T.H.).

We thank Y. A. Chabbert and Susan Michelson for their helpful criticism. Wild resistant streptococcal strains were a gift from R. Minck (Université Louis Pasteur, Strasbourg), H. Dabernat (Faculté de Médecine, Toulouse), J. F. Acar (Hôpital St. Joseph, Paris), Annie Buu-Hoi (Hôpital Broussais, Paris) and A. L. Courtieu (C. H. R. Hôtel-Dieu, Nantes). We thank P. Lemoine for expert photographic assistance and Odette Rouelland for secretarial assistance.

REFERENCES

1. Eickhoff, T.C., Klein, J.O., Daly, A.K., Ingall, D. and Finland, M. (1964): Neonatal sepsis and other infections due to group B beta-hemolytic streptococci. *N. Engl. J. Med.*, *271*, 1221.
2. Dixon, J.M.S. and Lipinski, A.E. (1974): Infections with beta-hemolytic streptococcus resistant to lincomycin and erythromycin and observations on zonal-pattern resistance to lincomycin. *J. Infect. Dis.*, *130*, 351.

3. Kahlmeter, G. and Kamme, K. (1972): Tetracycline-resistant group A streptococci and pneumococci. *Scand. J. Infect. Dis., 4*, 193.
4. Mitsuhashi, S., Inoue, M., Fuse, A., Kaneko, Y. and Oba, T. (1974): Drug resistance in *Streptococcus pyogenes. Jpn. J. Microbiol., 18*, 98.
5. Clewell, D.B. and Franke, A.E. (1974): Characterization of a plasmid determining resistance to erythromycin, lincomycin, and vernamycin Bα in a strain of *Streptococcus pyogenes. Antimicrob. Agents Chemother., 5*, 534.
6. Horodniceanu, T., Bouanchaud, D.H., Bieth, G. and Chabbert, Y.A. (1976): R plasmids in *Streptococcus agalactiae* (group B). *Antimicrob. Agents Chemother., 10*, 795.
7. El-Solh, N., Bouanchaud, D.H., Horodniceanu, T., Roussel, A. and Chabbert, Y.A. (1978): Molecular studies and possible relatedness between R plasmids from groups B and D streptococci. *Antimicrob. Agents Chemother., 14*, 19.
8. Malke, H. (1979): Conjugal transfer of plasmids determining resistance to macrolides, lincosamides and streptogramin-B type antibiotics among groups A, B, D and H streptococci. *Fems Microbiol. Lett., 5*, 335.
9. Hershfield, V. (1979): Plasmids mediating multiple drug resistance in group B streptococcus: transferability and molecular properties. *Plasmid, 2*, 137.
10. Horodniceanu, T., Bougueleret, L., El-Solh, N., Bouanchaud, D.H. and Chabbert, Y.A. (1979): Conjugative R plasmids in *Streptococcus agalactiae* (group B). *Plasmid, 2*, 197.
11. Fuller, A.T. (1938): The formamide method for the extraction of polysaccharides from hemolytic streptococci. *Br. J. Exp. Pathol., 19*, 130.
12. Drapkin, M.S., Karchmer, A.W. and Moellering, R.C., Jr (1976): Bacteremic infections due to clindamycin-resistant streptococci. *J. Am. Med. Assoc., 236*, 263.
13. Jacobs, M.R., Mithal, Y., Robins-Browne, R.M., Gaspar, M.N. and Koornhof, H.J. (1979): Antimicrobial susceptibility testing of pneumococci: determination of Kirby-Bauer break-points for penicillin G, erythromycin, clindamycin, tetracycline, chloramphenicol, and rifampin. *Antimicrob. Agents Chemother., 16*, 190.
14. Weisblum, B. (1974): Altered methylation of ribosomal ribonucleic acid in erythromycin-resistant *Staphylococcus aureus.* In: *Microbiology-1974*, p. 199. Editor: D. Schlessinger. American Society for Microbiology, Washington, D.C.
15. Weisblum, B., Holder, S.B. and Halling, S.M. (1979): Deoxyribonucleic acid sequence common to staphylococcal and streptococcal plasmids which specify erythromycin resistance. *J. Bacteriol., 138*, 990.
16. Rousset, A., Levy, A. and Minck, R. (1977): Group B streptococci: serotyping data and susceptibility to antibiotics. *Ann. Microbiol., 128B*, 339.
17. Sanders, E., Foster, M. T. and Scott, D. (1968): Group A beta-hemolytic streptococci resistant to erythromycin and lincomycin. *N. Engl. J. Med., 278*, 538.
18. Dixon, J.M.S. (1968): Group A streptococcus resistant to erythromycin and lincomycin. *Can. Med. Assoc. J., 99*, 1093.
19. Nakae, M., Murai, T., Kaneko, Y. and Mitsuhashi, S. (1977): Drug resistance in *Streptococcus pyogenes* isolated in Japan (1974–75). *Antimicrob. Agents Chemother., 12*, 427.
20. Hyder, S.B. and Streitfeld, M.M. (1973): Inducible and constitutive resistance to macrolide antibiotics and lincomycin in clinically isolated strains of *Streptococcus pyogenes. Antimicrob. Agents Chemother., 4*, 327.
21. Yagi, Y., Franke, A.E. and Clewell, D.B. (1975): Plasmid-determined resistance to erythromycin: comparison of strains of *Streptococcus faecalis* and *Streptococcus pyogenes* with regard to plasmid homology and resistance inducibility. *Antimicrob. Agents Chemother., 7*, 871.
22. Courvalin, P.M., Carlier, C. and Chabbert, Y.A. (1972): Plasmid-linked tetracycline and erythromycin resistance in group D streptococcus. *Ann. Inst. Pasteur, Paris, 123*, 755.
23. Clewell, D.B., Yagi, Y., Dunny, G.M. and Schultz, S.K. (1974): Characterization of three plasmid deoxyribonucleic acid molecules in a strain of *Streptococcus faecalis*: identification of a plasmid determining erythromycin resistance. *J. Bacteriol., 117*, 283.

Methods for the identification of streptococci in a clinical bacteriological laboratory

C. E. Nord
National Bacteriological Laboratory, Stockholm, Sweden

INTRODUCTION

There is no generally accepted classification of streptococci (Table I). Pyogenic streptococci are β-haemolytic streptococci that possess a Lancefield polysaccharide group antigen. Most clinically important streptococci possess Lancefield group antigens A, B, C, D, G or F. Grouping of streptococci is important for the treatment and the prognosis of the patient.

Pyogenic streptococci of group A are most often called *Streptococcus pyogenes* (Table II). *S. pyogenes* is the usual bacterial causative agent of tonsillitis, and complications of *S. pyogenes* infection are rheumatic fever and acute glomerulonephritis.

Table I. Streptococci found in man.

Group	Common group name	Common species name	Lancefield group antigen	Haemolysis*
Pyogenic streptococci	Group A	*S. pyogenes*	A	β
	Group B	*S. agalactiae*	B	β
	Group C	*S. equisimilis*	C	β
	Group G	–	G	β
Enterococci	Faecal	*S. faecalis*	D	$-/\beta$
	Faecal	*S. faecium*	D	v
Alpha-haemolytic streptococci		*S. sanguis*	–(H)	$\alpha(\beta, -)$
		S. mitior	–(O, K, M)	$\alpha(\beta, -)$
Non-haemolytic streptococci		*S. salivarius*	–(K)	–
		S. mutans	–(E)	–
		S. bovis	D	–
		S. milleri	–(A, C, F, G)	$-(\beta, \alpha)$

* Less common haemolytic appearances in parentheses; v = various; $-$ = non-haemolytic

Table II. Biochemical characteristics of β-haemolytic human streptococci.

Characteristic	S.pyogenes	S.agalactiae	S.equisimilis	S.faecalis	S.faecium	Group G
Growth 40 % bile	–	+	–	+	+	–
Growth 45 °C	–	–	–	+	+	–
Esculin hydrolysis	V	–	V	+	+	V
Hippurate hydrolysis	–	+	–	+	V	–
Acid from sorbitol	–	–	–	+	V	–
Acid from mannitol	–	–	–	+	V	–
Growth on 0.4 % tellurite	–	V	–	+	–	–

+ = 90 % positive strains
– = 90 % negative strains
V = some strains positive, some strains negative

The β-haemolytic members of groups C and G are usually referred to as groups C and G. Streptococci found in man and belonging to group C are similar to *S. equisimilis* (Table II). However, no species name for such streptococci has been proposed. Streptococci of groups C and G are isolated from different infections such as tonsillitis, septicaemia and endocarditis.

Group B streptococci found in man are called *S. agalactiae* (Table II) and are probably distinct from bovine group B streptococci.

S. agalactiae can cause neonatal meningitis and septicaemia but also urinary tract infections and wound infections in adults.

Faecal streptococci, often called enterococci (*S. faecalis*, *S. faecium*, Table II) possess group D antigen and can be recovered from serious infections such as endocarditis and septicaemia but also from urinary tract infections.

There also exist streptococci found in man which possess several Lancefield polysaccharide antigens. *S. milleri*, for example, can possess antigens of groups A, C, F or G.

Viridans streptococci are less well defined but can be divided into two groups, α-haemolytic and non-haemolytic. These streptococci often possess no Lancefield antigens. Identification of this group, therefore, requires the performance of biochemical and physiological tests in addition to Lancefield grouping.

Among the α-haemolytic streptococci, two rather well-recognized species are described: *S. sanguis* and *S. mitior* (Table III). *S. sanguis* has some typical biochemical characteristics such as hydrolysis of arginine and esculin, formation of dextran and acid production from trehalose and inulin. About 50 % of the *S. sanguis* strains have the group H antigen. This species is often isolated from endocarditis and septicaemia but has also been recovered from purulent infections.

Alpha-haemolytic streptococci which are not biochemically active, especially with respect to the characteristics mentioned above, belong to the species *S. mitior*. Some strains have the group antigen K, M or O. *S. mitior* can also be isolated from systemic diseases such as endocarditis and septicaemia.

A third group of α-haemolytic streptococci, which form dextran but do not ferment arginine or esculin, should be recognized because its pathogenic potential resembles that of *S. sanguis* rather than that of *S. mitior*. The remainder of the α-haemolytic

Table III. Biochemical characteristics of α-haemolytic streptococci found in man.

Characteristic	S. mitior	S. sanguis
Growth 40 % bile	V	V
Growth 45 °C	−	V
Esculin hydrolysis	−	+
Arginine hydrolysis	−	+
Acetoin from glucose	V	−
Acid from trehalose	−	+
sorbitol	−	−
mannitol	−	−
raffinose	V	V
inulin	−	+
Levan from sucrose	−	−
Dextran from sucrose	V	+

Symbols as for Table II

streptococci – dextran-negative strains that hydrolyse either arginine or esculin but not both – cannot yet be classified further by the recommended tests.

Four species of predominantly non-haemolytic streptococci can be recognized in material from human sources: *S. milleri, S. mutans, S. salivarius* and *S. bovis* (Table IV).

S. milleri is a species which includes biochemically similar strains with a characteristic cell wall composition. The majority of the strains are non-haemolytic and possess no group antigen. However, some strains are β-haemolytic and some α-haemolytic. About 25 % of the haemolytic strains have the group antigens A, C, G or F. Haemolysis and group antigens occur independently but there is a tendency for β-haemolytic isolates to be groupable more often than α- or non-haemolytic strains.

Table IV. Biochemical characteristics of non-haemolytic streptococci found in man.

Characteristic	S. salivarius	S. mutans	S. milleri	S. bovis I	S. bovis II
Growth 40 % bile	−	−	−	+	+
Growth 45 °C	−	V	−	V	V
Esculin hydrolysis	+	+	V	+	+
Arginine hydrolysis	−	−	V	−	−
Acetoin from glucose	V	V	V	+	+
Acid from trehalose	V	+	V	+	+
sorbitol	−	+	−	−	−
mannitol	−	+	−	+	V
raffinose	+	+	−	+	+
inulin	+	+	−	+	V
Levan from sucrose	+	−	−	−	−
Dextran from sucrose	−	+	−	+	−

Symbols as for Table II

134

The remarkable ability of *S. milleri* to form localized abscesses in different organs such as the central nervous system, thorax and abdomen separate this species from other human streptococci.

S. mutans forms a well-defined species. The production of acid from mannitol and sorbitol and the formation of dextran from sucrose are typical biochemical characteristics. Lancefield group antigen E can be found in some strains. *S. mutans* has been implicated in dental caries and endocarditis.

S. salivarius is characterized by the ability to produce levan from sucrose and therefore gives rise to large mucoid colonies on sucrose agar. About half of the strains possess Lancefield group K antigen. *S. salivarius* is seldom isolated from human infections and is therefore considered to be non-pathogenic.

S. bovis is divided into two biotypes. Biotype I strains form dextran, hydrolyse starch and usually ferment mannitol and inulin, while biotype II strains are dextran- and starch-negative and much less often ferment mannitol and inulin. *S. bovis* possesses the group D antigen and is often called non-enterococcal group D streptococci. The dextran-forming *S. bovis* biotype I is often associated with endocarditis.

During the last few years there has been an increased interest in developing new techniques suitable for clinical laboratories and in improving classical streptococcal identification.

SEROLOGICAL CLASSIFICATION OF β-HAEMOLYTIC STREPTOCOCCI

Extraction of streptococcal group antigen

The definite identification of β-haemolytic streptococci is made by demonstration of the group-specific carbohydrate antigen that is extracted from the streptococci. Different methods for extracting the group carbohydrate antigen have been published: the hot-acid method of Lancefield [1], the hot-formamide method of Fuller [2], the autoclave method of Rantz and Randall [3] and the enzyme method of Maxted [4]. A method using proteolytic enzymes such as pronase B has also been described [5]. Recently, El Kholy and co-workers [6] have used nitrous acid extraction for liberation of streptococcal group antigen. All extraction techniques have been reported to work satisfactorily. Lancefield's HCl extraction and Fuller's formamide extraction are still the reference methods.

Identification of streptococcal group antigen

After extraction of the streptococcal antigen, serological techniques are used for antigen identification. Three different methods are mostly used: precipitin capillary tube test [1], immunodiffusion [7] and countercurrent immunoelectrophoresis [8]. The immunodiffusion method is slower than the capillary tube test but more accurate because lines of identity can be observed between known control antigen and the antigens under investigation. Countercurrent immunoelectrophoresis has the advantage that small volumes and weaker reagents can be used and the result is obtained fairly rapidly.

However, these methods are laborious and time-consuming in routine clinical bacteriological work since growth of bacteria, extraction and identification are required.

Serological grouping techniques using whole streptococcal cells

The fluorescent antibody (FA) method and agglutination techniques are examples of streptococcal grouping methods in which whole bacterial cells are used.

Immunofluorescent techniques. Immunofluorescence is used in many laboratories for identification of group A streptococci. It can also be used for identification of β-haemolytic streptococci of groups B, C, D and G. The FA method is accurate when good specific antisera are used [9]. It is possible with this method to identify group A streptococci directly from throat swabs.

Co-agglutination test. Streptococcal antibodies are bound to the protein A component of the cell wall of staphylococci. When a sample containing streptococci is mixed with its corresponding antibody reagent, the specific antigens on the streptococcal cell surface bind to the corresponding specific antibodies. The resulting co-agglutination is visible within one minute [10].

Studies have shown that when co-agglutination is used, more than 90 % of group A, B, C and G streptococci can be identified as compared with the Lancefield grouping technique [11].

Latex agglutination test. In the latex agglutination test, a simple enzymatic extraction procedure is used. Streptococcal antigen is identified using polystyrene latex particles which have been coated with group-specific antibodies. These particles agglutinate strongly in the presence of homologous antigen and remain in smooth suspension in the presence of heterologous antigen.

Comparative investigations between the latex agglutination method and the Lancefield precipitation test have shown that the agglutination test is a rapid and reliable test for identification of most streptococcal strains belonging to groups A, B, C, F and G [12].

IDENTIFICATION OF β-, α- AND NON-HAEMOLYTIC STREPTOCOCCI BY BIOCHEMICAL TESTS

The clinically most important β-haemolytic streptococci – groups A, C and G – cannot be identified by biochemical and physiological tests only. However, screening techniques are available. These presumptive tests cannot replace serological methods for accurate identification of β-haemolytic streptococci but they are useful for preliminary identification [13]. The following tests can usefully be carried out in the clinical laboratory: determination of haemolysis, susceptibility to bacitracin, hippurate hydrolysis, bile-esculin hydrolysis and tolerance to 65 % NaCl (Table V).

Susceptibility to bacitracin is a screening method for the presumptive identification of group A streptococci. The test is reliable if carried out correctly, i.e., with respect to concentration of bacitracin used and interpretation of zone sizes. Some strains belonging to groups C and G may be sensitive to bacitracin. Resistant group A strains are seldom recovered under defined test conditions. Beta-haemolytic streptococci that are susceptible to bacitracin, do not hydrolyse hippurate and fail to blacken bile-esculin medium, are presumptively streptococci of group A.

Beta-haemolytic streptococci that hydrolyse hippurate and fail to blacken bile-

Table V. Presumptive identification of clinically important streptococci.

Streptococci	Haemolysis	Bacitracin susceptibility	Hippurate hydrolysis	Bile-esculin hydrolysis	Tolerance to 6.5% NaCl
Group A	β	+	−	−	−
Group B	β	−	+	−	V
Group D enterococci	α, β, none	−	V	+	+
Group D non-enterococci	α, none	−	−	+	−
Non group A, B or D	β	−	−	−	−
Viridans	α, none	V	−	−	−

esculin medium are presumptively streptococci of group B. Bacitracin-susceptible streptococci which hydrolyse hippurate are also presumptive group B streptococci. Beta-haemolytic streptococci that are resistant to bacitracin, fail to hydrolyse hippurate or blacken bile-esculin medium and are not tolerant to 6.5% NaCl are streptococci not belonging to groups A, B or D.

Bile-esculin positive strains are group D streptococci and 6.5% NaCl-tolerant group D streptococci are enterococci. Enterococci can be α-, β- or non-haemolytic. Some strains of enterococci hydrolyse hippurate. Enterococci are hippurate-positive, bile-esculin-positive and salt-tolerant. Most streptococci of group D are resistant to bacitracin.

Alpha-haemolytic streptococci which fail to blacken bile-esculin medium, do not hydrolyse hippurate and are not 6.5% NaCl-tolerant are viridans streptococci. Some viridans streptococci are sensitive to bacitracin. For further identification of viridans streptococci, the following tests can be carried out: production of acid from mannitol, inulin and raffinose, production of acetoin from glucose, hydrolysis of arginine and esculin, growth in 40% bile and dextran and levan formation from sucrose.

Test kits

Diagnostic biochemical kits for grouping streptococci are also available. One kit, API-STREP®, designed to identify β-haemolytic streptococci groupw A, B, C, D and G, has been evaluated. The system identifies streptococci of groups A, B and D acceptably but not those of groups C and G.

CONCLUSIONS

A variety of methods is available for the identification of human streptococci in the clinical bacteriological laboratory. The choice of the most suitable methods should be based on evaluation of cost, time, availability of equipment and reagents, technical experience and the number of tests performed routinely. Consideration should also be given to the reproducibility of the method for the identification of all human streptococci of clinical importance.

REFERENCES

1. Lancefield, R. (1933): A serological differentiation of human and other groups of hemolytic streptococci. *J. Exp. Med., 57*, 581.
2. Fuller, A.T. (1938): The formamide method for the extraction of polysaccharides from haemolytic streptococci. *Br. J. Exp. Pathol., 19*, 130.
3. Rantz, L.A. and Randall, E. (1955): Use of autoclaved extracts of hemolytic streptococci for serological grouping. *Stanford Med. Bull., 13*, 290.
4. Maxted, W.R. (1948): Preparation of extracts for Lancefield grouping. *Lancet, II*, 255.
5. Ederer, G.M., Herrmann, M.M., Bruce, R., Matsen, J.M. and Chapman, S.S. (1972): Rapid extraction method with pronase B for grouping beta-haemolytic streptococci. *Appl. Microbiol., 23*, 285.
6. El Kholy, A., Facklam, R., Sabri, G. and Rotta, J. (1978): Serological identification of group A streptococci from throat scrapings before culture. *J. Clin. Microbiol., 8*, 725.
7. Holm, S.E. (1965): A modified gel chamber technique for immuno-diffusion analysis. *Int. Arch. Allergy Appl. Immunol., 26*, 34.
8. Wadström, T., Nord, C.E., Lindberg, A.A. and Möllby, R. (1974): Rapid grouping of streptococci by immunoelectroosmophoresis. *Med. Microbiol. Immunol., 159*, 191.
9. Moody, M.D., Ellis, E.C. and Updyke, E.L. (1958): Staining bacterial smears with fluorescent antibody. IV. Grouping streptococci with fluorescent antibody. *J. Bacteriol., 75*, 553.
10. Christensen, P., Kahlmeter, G., Jonsson, S. and Kronvall, G. (1973): New method for the serological grouping of streptococci with specific antibodies adsorbed to protein A-containing staphylococci. *Infect. Immun., 7*, 881.
11. Hahn, G. and Nyberg, I. (1976): Identification of streptococcal groups A, B, C and G by slide co-agglutination of antibody-sensitized protein A containing staphylococci. *J. Clin. Microbiol., 4*, 99.
12. Facklam, R.R., Cooksey, R.C. and Wortham, E.C. (1979): Evaluation of commercial latex agglutination reagents for grouping streptococci. *J. Clin. Microbiol., 10*, 641.
13. Facklam, R.R., Padula, J.F., Thacker, L.G., Wortham, E.C. and Sconyers, B.J. (1974): Presumptive identification of group A, B and D streptococci. *Appl. Microbiol., 27*, 107.

Discussion

J. Rotta (Prague, Czechoslovakia): With regard to the bile-esculin test, of course in group B this test is negative in about 10 % of the cases.

Secondly, the method of mechanical disintegration used for the extraction of the polysaccharides, as presented by Dr Nord, is restricted to group D only. I wonder whether it is possible to liberate the group A, C and G polysaccharides by mechanical disintegration as well.

C. E. Nord: Mechanical disintegration should only be used for extraction of group D antigen.

R. R. Facklam (Atlanta, U.S.A.): What was the reference method to which the agglutination reactions were compared?

C. E. Nord: The Lancefield precipitin test was the reference method.

S. A. Waitkins (Liverpool, England): Were any viridans group Ds used for the Streptex® method?

C. E. Nord: No.

S. A. Waitkins: If not, there might be a different result when API® is used.

R. R. Facklam: Can you clarify that you are talking about different commercial methods?

S. A. Waitkins: When Dr Nord used API-STREP® he obtained only one of three group Ds which makes this method look bad. In fact, however, if we look for viridans group Ds – *Streptococcus bovis*, for example – it will not be found with Streptex®, which is a serological method, but it will certainly be found with API®. It was slightly misleading to make API® appear bad on the group Ds. I agree about group C and G and think that API® is not so good for those organisms. It is a fairly good test on the group Ds.

R. R. Facklam: Basically, the API® system will work to identify non-β-haemolytic streptococcal strains?

S. A. Waitkins: No, β-haemolytic *Streptococcus faecalis* are also obtained.

C. E. Nord: The API-STREP® system is designed for the identification of haemolytic streptococci. In our investigation it was not possible to differentiate between streptococci group C and G, using only the API-STREP® system.

G. Laurell (Uppsala, Sweden): In some Swedish laboratories nucleatase is used in combination with bacitracin, the nucleatase stimulating the haemolysis. The combination of these two compounds seems to be safer for group A than bacitracin alone. Dr Nord did not mention that procedure, and I do not know whether it is commonly used in other countries, or not used anywhere other than Uppsala.

C. E. Nord: We have been studying the use of that combination and I agree about it.

G. Laurell: Is the combination used in the United States?

R. R. Facklam: No. I have my own system of presumptive identification which is rather involved. I would prefer not to go into any detail about it now. There is an article in the *Journal of Clinical Microbiology*, June 1979 issue, stating that it is possible to classify 97 % of the β-haemolytic streptococci into three categories. However, these rapid grouping procedures we have been discussing today should replace the presumptive identification schemes. In my opinion they provide accurate identification and are as rapid if not more so than presumptive testing.

Comparative study of different methods of grouping streptococci A, B, C, D and G

F. Tessier and G.-L. Daguet
Central Bacteriological-Virological Service, St Antoine Hospital, Paris, France

INTRODUCTION

The identification of a streptococcus cannot be complete without determining the group to which it belongs. Grouping is usually done by the Lancefield method, still regarded as the reference method. Whether the Lancefield method or Fuller's technique are considered, both are long and require a great deal of care, making rapid results impossible. Today, group identification often makes it possible to assess the pathogenic potential of the streptococcus isolated and to predict its sensitivity to antibiotics. For this reason we decided to investigate quicker grouping techniques: *immunological,* by slide agglutination, after adsorption of antibody either to protein A of staphylococci (Phadebact® Streptococcus Test (Pharmacia)) [1] or to polystyrene latex particles (Streptex® (Wellcome)) [2]; and *biochemical,* based on the detection of enzymatic activities specific to the main serological groups (API-STREP® (Api-System)) [3].

MATERIAL AND METHODS

Isolates

Two hundred and forty-five strains of streptococci (A: 80, B: 70, C: 6, D: 58, G: 31) isolated from various samples between January, 1978, and January, 1979, in the bacteriological department of the St Antoine Hospital were grouped by the following methods:
- API-STREP®: 245,
- Phadebact® Streptococcus Test: 187 (there is no anti-D reagent in this method),
- Streptex® reagents: 154 (available only since September 1978).
All strains were checked by the Fuller method.

Culture

From a culture on either blood agar plate, allowing appreciation of haemolytic characteristics, or on ascites agar plate, allowing examination by oblique trans-illumination, a subculture was made in Todd Hewitt broth for the agglutination

141

technique using the Phadebact® Streptococcus Test system, in buffered dextrose broth for agglutination by Streptex® reagents and for preparation of the antigenic extract by the Fuller method or in API-STREP® medium for preparation of the API strip.

In the case of streptococci of group D, detected by blackening of bile-esculin medium, certain cultural or biochemical characteristics have been investigated using the usual techniques – tolerance to 6.5 % NaCl, black colonies on potassium tellurite, fermentation of mannitol, sorbitol and raffinose and hydrolysis of starch. These allow classification into four species: *faecalis, faecium, durans* and *bovis* [4–6].

Group determination

We followed the directions for use provided with the reagents for each of the techniques used:
Phadebact® Streptococcus Test. Observation of agglutination on a glass slide, directly from the Todd Hewitt broth. This is possible as soon as there is sufficient culture growth.
Streptex®. Observation of agglutination on a glass slide with the supernatant obtained from a pellet of bacteria treated with pronase for one hour at 56 °C.
API-STREP®. Inoculation of 10 characteristic API strips from a centrifugation pellet of the 24-hour culture, washed and resuspended in 1 ml of distilled water. The biochemical characteristics are revealed after three hours' incubation at 37 °C.
Fuller method [7].

Alternative techniques

The above-mentioned techniques were carried out starting from cultures in broth and are those for which results are reported. Some later modifications have been suggested which, as well as simplifying the techniques, allow results to be obtained more rapidly:
– agglutination with Phadebact® Streptococcus Test reagents from colonies taken from the agar plate [8],
– agglutination with Streptex® reagents from colonies treated with pronase,
– agglutination with Phadebact® Streptococcus Test reagents directly on the plate [9].
Our personal experience with this latter technique will be brought up in the discussion.

RESULTS AND COMMENTS

Results are expressed as the number of positive responses compared with the number of strains tested.

Streptococci of group A: 80 strains

Fuller: 79/80
Phadebact® Streptococcus Test: 80/80
API-STREP®: 78/80
Streptex®: 48/48
Correlation with immunological methods was excellent. A single strain was ne-

gative in the Fuller method but not in the others. With the API strips there was a minor divergence in one case where β-galactosidase was found to be slightly positive instead of negative while in another the enzymatic characteristics obtained suggested a nongroupable streptococcus. The strain was, moreover, nonhaemolytic and clouded the broth. The mannitol + characteristic was found twice, indicating a biochemical variant.

Streptococci of group B: 70 strains

Fuller: 70/70
Phadebact® Streptococcus Test: 70/70
API-STREP®: 70/70
Streptex®: 44/44
In all the tests used the correlation was excellent. However, among the other biochemical characteristics which were uniformly positive, β-glucuronidase was negative in half of all cases (35/70). In fact, if the reagents are homogenized in the well the coloration will appear, proving the test result positive.

Streptococci of groups C and G: 37 strains

Groups C and G were combined because of the small number of C isolated and the great similarity of the enzymatic characteristics of the two groups.
Fuller: C 6/6; G 31/31
Phadebact® Streptococcus Test: C 6/6; G 31/31
API-STREP®: C + G 34/37
Streptex®: C 3/3; G 24/24
Correlation with the immunological methods was excellent. Correlation with the enzymatic characteristics was good if streptococci of groups C and G were not differentiated. With the streptococci of group G results were good but with those of group C results were less consistent: three strains were found to be esculin + and β-glucuronidase −, thus classing them among the nongroupables.

Streptococci of group D: 58 strains

A distinction has to be made between the immunological methods (Fuller, Streptex®) and the biochemical methods (API-STREP®, usual media), allowing the variety of streptococcus D to be determined.

1. Immunological methods
Fuller: 56/58
Streptex®: 35/35
There were two negative strains with the Fuller method (one *faecalis*, one *bovis*) but excellent correlation with Streptex® reagents. It should be noted, however, that while all streptococci of group D studied gave an agglutination with Streptex® D reagent, some nongroupable streptococci were able to give agglutination with the D reagent. Grouping with Phadebact® Streptococcus Test reagents showed that certain D *faecalis* strains could give agglutination with the C reagent. These observations led us to inoculate a bile-esculin medium, for all streptococci studied, on a routine basis.

2. Biochemical methods

Culture on bile-esculin and in broth at 45 °C was positive for all 58 isolates. Biochemical characteristics showed an excellent correlation between the two types of method and permitted the identification of strains as follows:

S. faecalis   37 (API-STREP®); 37 (classical techniques)
S. faecium   5 (API-STREP®); 5 (classical techniques)
S. durans   2 (API-STREP®); 2 (classical techniques)
S. bovis   8 (API-STREP®); 8 (classical techniques)
?   6 (API-STREP®); 6 (classical techniques)

Six strains grouped as D presented biochemical characteristics which did not allow them to be categorized with certainty in any given species. There was a lack of biochemical characteristics, both in conventional media and on API strips.

DISCUSSION AND CONCLUSIONS

Taking into account the remarks above, the results obtained by the different methods are very satisfactory. The Phadebact® Streptococcus Test system has the advantage of being simple and rapid and offers excellent reliability [8, 10, 11]. When applied to the screening of streptococci B in the perinatal period it can give a response in three to four hours [12].

More recently, we adapted Edwards' and Larson's technique [9], which consists of depositing a drop of serum on the suspected colony, to streptococci B [2]. After rocking the plate slightly, agglutination takes place and is visible under a magnifying glass within a few seconds of mixing. It is generally very clear and specific. It can also be obtained on agar, after removal of the colony for subculture. Out of 50 streptococci B tested, we obtained 49 definitely positive reactions and one doubtful. Twenty streptococci which did not belong to group B gave no agglutination. Diagnosis of streptococcus B can, therefore, be established from primary culture in a solid medium and without any delay.

We checked agglutination for each strain after incubation in Todd Hewitt broth. Spots on ascites agar plates showed no agglutination with the reagents A and C and mostly not with G. (Slight agglutination was noticed in three cases with reagent G.)

The Streptex® reagents proved to be extremely reliable for streptococci of all groups including D. Theoretically, their use needs a heavy growth (usually 24 hours) and pronase extraction. Results are, therefore, less rapidly obtained. Extraction from colonies sampled from an agar plate is possible if the suspension is thick enough. We have tried extraction several times directly from blood culture broth and obtained the antigenic group within two hours. These alternative techniques cut down the time needed to obtain results.

The enzymatic test on API strip is also reliable, if the instructions for use are followed. It also needs a 24-hour culture but permits the identification of species within group D as well as the detection of certain biochemical types in the other groups which could be of epidemiological interest. At the moment these techniques can be applied to certain nongroupable streptococci.

The choice of one or the other of these techniques is up to the user, according to his needs and possibilities. Helped by the appearance of the cultures, haemolysis characteristics and a simple biochemical test for orientation or control, they permit the rapid and reliable grouping of streptococci isolated.

144

SUMMARY

Two hundred and forty-five strains of streptococci isolated from various samples obtained in the hospital were grouped by different methods: *immunological*, by slide agglutination using antibodies adsorbed either to protein A of staphylococci (Phadebact® Streptococcus Test) or to polystyrene latex particles (Streptex®); and *biochemical*, by detection of enzymatic activities specific to the main serological groups (API-STREP®). The results were compared with those obtained by the classical Fuller method.

Correlation was good, even excellent, for all strains tested. Culture on bile-esculin medium enables correction to be made for false positive reactions obtained with some nongroupable and D streptococci.

The routine introduction of these new techniques should allow rapid identification of streptococci isolated and a reduction in work, making it possible to extend grouping to a greater number of strains.

REFERENCES

1. Christensen, P., Kahlmeter, G., Jonsson, S. and Kronvall, G. (1973): A new method for the serological grouping of streptococci with specific antibodies absorbed to protein A-containing staphylococci. *Infect. Immun.*, 7, 881.
2. Farrar, J.L. and Paull, A. (1979): Streptococcal disease in man and animals. In: *Pathogenic Streptococci*. Editor: M.T. Parker. Reedbooks, Chertsey.
3. Waitkins, S. A. (1976): Identification of streptococci using API-ZYM system. In: *Second International Symposium on Rapid Methods and Automatization in Microbiology, Cambridge, September 1976*, p. 67. Learned Information Europe.
4. Bergey, S. (1974): *Manual of Determinative Bacteriology*, 8th ed. Editors: R.E. Buchanan and N.E. Gibbons. Williams and Wilkins Company, Baltimore.
5. Hardy, M.A., Dalton, H.P. and Allison, M.J. (1978): Laboratory identification and epidemiology of streptococcal hospital isolates. *J. Clin. Microbiol.*, 8, 534.
6. Wechsler, B., Levantis, S., Herreman, G. and Godeau, P. (1978): Association cancer digestif et endocardite subaiguë. *Nouv. Presse Méd.*, 7, 2845.
7. Fuller, A.T. (1938): The formamid method for the extraction of polysaccharide from beta-hemolytic streptococci. *Br. J. Pathol.*, 19, 131.
8. Pillot, J., Lebrun, L. and de Maneville, M. (1977): Progrès dans l'identification des streptocoques pathogènes. Le groupage par co-agglutination. *Nouv. Presse Méd.*, 6, 1822.
9. Edwards, E.A. and Larson, G.L. (1974): A new method of grouping beta-haemolytic streptococci directly on sheep blood agar plates by co-agglutination of specifically sensitized protein A-containing staphylococci. *Appl. Microbiol.*, 28, 972.
10. Deschamps, C., Ortenberg, M. and Perol, Y. (1976): Identification immunologique de groupe des streptocoques par une technique rapide d'agglutination sur lame. *Méd. Mal. Inf.*, 6, 466.
11. Szilagyi, G., Mayer, E. and Eidelman, A.I. (1978): Rapid isolation and identification of group B streptococci from selective broth medium by slide co-agglutination test. *J. Clin. Microbiol.*, 8, 410.
12. Tessier, F., Colau, J.C., Bouillié, J., Le Lorier, G. and Daguet, G.-L. (1977): Risque infectieux néonatal à streptocoque B. *J. Gynecol. Obstet. Biol. Reprod.*, 6, 239.

Discussion

J. Rotta (Prague, Czechoslovakia): Has Professor Daguet any information about the types of group B streptococci which he has been able to grow directly on plates? Have there been different polysaccharide types or only one?

G. L. Daguet: Unfortunately I do not know about the types grown in that way.

A.-S. Malmborg (Stockholm, Sweden): Which medium was used for direct agglutination on the plates? I think is was serum-free.

G. L. Daguet: It was an ascitic fluid aga4 plate.

C. W. Woodmansea (Leeds, England): Has Professor Daguet tried Phadebact® Streptococcus Test co-agglutination direct on the blood culture, in the same way as he did with Streptex®?

G. L. Daguet: No, that was not tried.

Practical evaluation of methods for the routine laboratory identification of β-haemolytic streptococci

H. Arvilommi
Public Health Laboratory, Jyväskylä, Finland

We have looked at the question of serological grouping of streptococci from the point of view of smaller laboratories which usually have to rely on commercial reagents. In an earlier study [1] we compared rapid grouping methods and found that, compared to precipitation, a modification of the co-agglutination technique (COA) [2] gave identical results in 98%, the fluorescent antibody test in 89% and counterimmunoelectrophoresis in 96% of cases. Subsequently we adopted COA

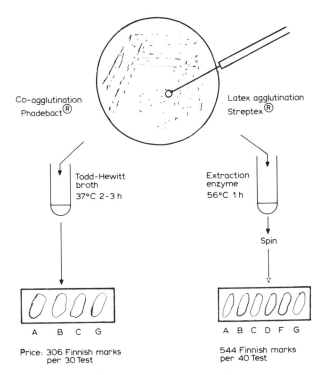

Figure. Comparison of co-agglutination and latex agglutination techniques.

as the routine method for grouping β-haemolytic streptococci in our laboratory. I report here on the performance of COA in the hands of inexperienced laboratory workers and compare COA to a novel polystyrene latex agglutination method (LA) [3], now also commercially available.

MATERIALS AND METHODS

Strains of β-haemolytic streptococci isolated from clinical specimens subcultured on blood agar plates were used in the study. The modification of COA [1] consisted of three-hour broth incubation followed by slide co-agglutination (Phadebact® Streptococcus Test, Pharmacia Diagnostics AB, Uppsala, Sweden). A trial of direct co-agglutination was made by smearing colonies from the plates directly onto the slides, followed by drops of the COA reagents. The latex agglutination test (Streptex®, Wellcome Reagents Ltd, Beckenham, England) was used according to the manufacturers' instructions, except that 0.3 instead of 0.4 ml of solution of extraction enzyme was used (Fig.).

RESULTS AND DISCUSSION

The strains studied in this investigation were originally grouped in our routine daily diagnostic work. The grouping is usually performed by several technicians and, as it happened, temporary workers were also involved in the grouping during the period of strain collection. It was, therefore, of interest to evaluate the per-

Table I. Slide co-agglutination directly from colonies.

Group	Correct/Number tested	Percentage
A	5/23	22
B	29/40	72
C	1/ 5	20
D	1/ 7	14
Totals	36/75	48

Table II. Comparison of co-agglutination and latex agglutination in grouping of β-streptococci.

		Latex							
		A	B	C	D	F	G	NG	Totals
C	A	30							30
O	B		45		1			1	47
A	C			6					6
	G						13		13
	NG					1			1
Totals		30	45	6	1	1	13	1	97

formance of the test including interpretations by inexperienced staff. Of 95 strains, 92 (97 %) were correctly grouped, as confirmed later by COA and LA. The incorrect results were one C and one nongroupable (NG), which should both have been A, and one G instead of C.

Although earlier we had discouraging results with direct co-agglutination of colonies from blood agar plates, I decided to reinvestigate this modification as it is mentioned as an alternative in the manufacturers' directions for use. Table I shows that this simple test was unreliable, as only half of the strains were correctly grouped. Although B streptococci were easiest to group, a quarter were not correctly grouped. Recently, Carlson and McCarthy [4] have reported on a rapid modification which uses a short enzyme extraction of streptococcal colonies.

In the third phase of this study I compared COA with LA. The features of these tests are depicted in the Figure. Both tests need a rather heavy inoculum, a fact which necessitates subcultures from the original blood agar plates when growth is scanty. The latex test is more rapid but an additional centrifugation step preceding the agglutination erodes this advantage.

When the results of these tests are compared (Table II), it appears that in 95 strains out of 97 (98 %) the results were identical. Both the remaining strains were group B in COA but group D or nongroupable in the latex test.

It might be worth pointing out that the LA kit contains anti-D and anti-F reagents, which are not included in the COA kit. However, in our laboratory, in specimens from throats, wounds and exudates, the proportion of β-haemolytic streptococci nongroupable with COA, namely those of groups D and F and other even less common groups, is under 5%. In this study only one F strain was found by LA. The one D strain, which was B by COA, has not yet been confirmed.

In conclusion, three-hour COA and LA are very similar tests. Our experience with COA shows that it is very satisfactory as a routine method for grouping β-haemolytic streptococci in the clinical microbiology laboratory.

REFERENCES

1. Arvilommi, H., Uurasmaa, O. and Nurkkala, A. (1978): Rapid identification of group A, B, C and G beta-haemolytic streptococci by a modification of the co-agglutination technique. Comparison of results obtained by co-agglutination, fluorescent antibody test, counterimmunoelectrophoresis, and precipitation. *Acta Pathol. Microbiol. Scand. Sect. B*, 86, 107.
2. Christensen, P., Kahlmeter, G., Jonsson, S. and Kronvall, G. (1973): New method for the serological grouping of streptococci with specific antibodies adsorbed to protein A-containing staphylococci. *Infect. Immun.*, 7, 881.
3. Farrar, J.L. and Paull, A. (1979): Grouping of streptococci by means of latex particles coated with antibody. In: *Pathogenic Streptococci*, p. 256. Editor: M.T. Parker. Reedbooks; Chertsey.
4. Carlson, J.A. and McCarthy, L.R. (1979): Modified co-agglutination procedure for the serological grouping of streptococci. *J. Clin. Microbiol.*, 9, 329.

Discussion

I. Amirak (London, England): Dr Arvilommi agrees that the results with all methods studied were very similar and that he is using Phadebact® Streptococcus Test routinely in his laboratory. Is this because of cost? Are there any advantages or disadvantages of this and other methods?

H. Arvilommi: We adopted the co-agglutination method before the Latex method was marketed in Finland. After making our comparisons I see no reason to change our method.

I. Amirak: In our experience it gives an answer within three minutes very similar to the answer obtained after two hours' incubation. Is your experience similar? The former method is faster and easier for the technicians who find it preferable to have an answer in three minutes instead of waiting two hours and then having to start all over again.

J. Maeland (Trondheim, Norway): The fluorescent antibody test did not seem to work with the group B streptococci. Is there any explanation?

H. Arvilommi: I agree with Dr Facklam about this. We used commercial antisera and the only one which really worked was group A. None of the others was much good. The method is entirely dependent upon the quality of the antisera.

R. Facklam (Atlanta, U.S.A.): The B reagent will not work unless it is made from conjugate prepared from all the types as well as the group antigen. Dr Wilkinson, who worked in my laboratory several years ago, has published on this subject and has shown definitely that if a conjugate is prepared from pools made from the various types a very good group B reagent can be obtained. No commercially available group B antigen is made in this way.

J. Maeland: Does that mean that it is only antibody to the type-specific carbohydrate antigens that works and not the antibody to the group?

R. Facklam: Both antibodies are included in the FA reagent if made the way recommended by Dr Wilkinson, by making the FA conjugates from all four carbohydrate types plus the group B carbohydrate. The reason for this is that the type-carbohydrate antigen is on the outside of the group antigen. It is a capsular-type antigen and, unless there is sufficient group B-type antigen outside the group, the group B reagent will not react with those cells.

Experience and laboratory studies in grouping β-haemolytic streptococci with the co-agglutination grouping method, Phadebact® Streptococcus Test

I. D. Amirak, R. J. Williams*, B. C. J. Farrell and P. Noone
Microbiology Department, Royal Free Hospital, London, England

INTRODUCTION

Infections caused by β-haemolytic streptococci other than group A are increasingly being reported. The need to ascertain the Lancefield group of a β-haemolytic streptococcus has long been accepted and in clinical microbiological laboratories a rapid technique for grouping β-haemolytic streptococci with accuracy and simplicity is desirable.

In 1933, Lancefield [1] described a method of classifying β-haemolytic streptococci into groups, based on a precipitation reaction of the specific C-substance polysaccharide in the cell walls of these organisms. The significance of this classification was that the distribution of group antigens corresponded closely to the type of disease and the animal host from which the streptococci had been isolated. Agglutination reactions with the C substance had already been examined but had been found unsatisfactory. Lancefield concluded that there was insufficient concentration of group antigen at the cell surface to allow agglutination to occur reliably.

In 1943, Riskaer [2] showed that a slide agglutination method could be used to group streptococci if trypsinized cultures were used and if the antisera were adsorbed with cultures of the other groups which gave cross-reactions with unadsorbed sera. Rosendal [3] followed up this work, showing that agglutination was practicable as a routine method for grouping strains in groups A, C and G and that the results correlated well with Lancefield's precipitation method. The agglutination method has advantages over precipitation in that it requires less time, culture and antiserum.

In 1973, by modifying a method introduced by Kronvall [4] for typing pneumococci, Christensen and his colleagues [5] developed a co-agglutinating reagent consisting of formaldehyde and heat-treated (Cowan I) staphylococci coated with

---

* Present address: Department of Microbiology, The London Hospital, London, England

group-specific antibody to streptococci. These workers, using trypsin-treated cultures of β-haemolytic streptococci belonging to Lancefield groups A, B, C, D and G, found an absolute correlation between results obtained by co-agglutination and by Lancefield's method.

In 1976, the Phadebact® kit using the co-agglutination reaction for grouping β-haemolytic streptococci of groups A, B, C and G became commercially available. Since its introduction the kit has been extensively tested by various workers [6–8].

Prior to the availability of the Phadebact® kit, our laboratory used Lancefield's method (acid extraction) to group β-haemolytic streptococci. Although this method is accepted as a standard reference technique, it can be expensive in terms of technical time and volume of antiserum required. It may also produce results that are difficult to read, which often leads to the grouping of β-haemolytic streptococci into group A and non-group A [7]. Thus the introduction of a rapid, simple and reliable technique was viewed with interest.

RESULTS

Our studies with the Phadebact® kit fall into three sections.

Comparison of Phadebact® co-agglutination method with Lancefield method

Fifty consecutive clinical isolates of β-haemolytic streptococci were grouped by the Phadebact® method and by the Lancefield method as used in our laboratory (acid extraction). Beta-haemolytic streptococci isolated from clinical specimens were grouped by the Lancefield method by the staff of the routine laboratory and the results reported. The investigating team then grouped the same strains by the

Table I. Comparison of results of grouping 50 consecutive isolates of β-haemolytic streptococci by Lancefield and Phadebact® methods

Method	Lancefield group A B C D G	Total number of strains grouped	Total number of strains ungrouped
Lancefield	15 8 3 6 5	37	13
Phadebact®	17 10 8 * 5	40	10[+]

* Group D not detectable by Phadebact® method
[+] Includes 6 group D strains

Table II. Results of grouping 12 isolates of β-haemolytic streptococci by the Phadebact® method after four and 24 hours' incubation.

Time of incubation (hours)	Strain number											
	1	2	3	4	5	6	7	8	9	10	11	12
4	ng*	B	ng	A	G	B	ng	ng	B	G	ng	C
24	ng	B	A	A	G	B	C	C	B	G	ng	C

* ng = not groupable

152

Phadebact® co-agglutination method. The results are shown in Table I. The Phadebact® method grouped 40 out of 50 strains. Of the 10 which failed to group, six belonged to group D, for which there is no reagent in the Phadebact® kit. The Lancefield method grouped 37 out of 50 strains, and compared with Phadebact® there were failures in groups A, B and C. Four strains did not belong to groups A, B, C, D or G by Lancefield's method and were assumed to belong to other groups. The six group D strains all hydrolyzed esculin in bile-esculin agar.

Since it is desirable to obtain rapid results in a clinical laboratory and the Phadebact® kit has been recommended for use with four-hour broth cultures of β-haemolytic streptococci, the Phadebact® kit was tested in this part of the study on 12 strains grown for four hours and 24 hours. The results are shown in Table II. Seven out of 12 strains could be grouped after four hours' growth, thus allowing reports to be sent out within 24 hours of receiving the specimen. A further three strains could be grouped only after 24 hours' growth. Two strains were non-groupable.

Investigation of group D streptococci

Strains belonging to Lancefield's group D are not recognizable by the Phadebact® grouping method because there is no group D reagent in the kit. If group D strains are tested they may give rise to problems due to false-positive results with other group reagents. Following our initial study, described above, we grouped a further 100 strains of β-haemolytic streptococci by the Phadebact® method alone. Of these, two strains which hydrolyzed esculin and which were confirmed as group D strains by the Reference Laboratory gave positive co-agglutination reactions with the group G reagent. It was therefore decided to investigate the frequency of these false-positive reactions.

Fifty clinical isolates of β-haemolytic, α-haemolytic and non-haemolytic streptococci which were identified by positive esculin hydrolysis and Lancefield's grouping method as belonging to group D were tested by the Phadebact® method. The results are shown in Table III. Forty-four out of 50 strains gave no reaction with any of the reagents. One strain reacted positively with A, B and C reagents, three strains with G reagent and one strain each with B and C reagents, respectively.

As a result of the findings described here it was decided to replace the Lancefield method by the Phadebact® kit in our laboratory. To avoid confusing results which might arise with Group D strains a routine of inoculating a bile-esculin plate at the same time as the grouping broth was established. In addition, the strain was subcultured onto a blood agar plate containing 10% horse blood to check purity. On a few occasions, strains giving rise to reactions with more than one Phadebact® group reagent were found to be a mixture of two strains of β-haemolytic streptococci.

Table III. Results of testing of 50 group D streptococci by the Phadebact® method.

Number of group D strains grouped as				Number of strains failing to group	Percentage of tests with cross-reactions
A	B	C	G		
1	2	2	3	44	12

One strain gave a positive result with reagents for groups A, B and C

In one instance group A and group C and in another group A and group G strains were present. The two strains could be distinguished by colonial morphology on the blood agar plate and, when separated, each gave its individual group reaction. This finding has been reported elsewhere [6]. Anthony and colleagues [9] have also reported the presence of a second streptococcal group in lesions infected with Lancefield group A streptococci.

Effect of different commercial broth media on the Phadebact® co-agglutination method

Use of the Phadebact® co-agglutination technique for grouping β-haemolytic streptococci was continued satisfactorily for eight months with different batches of grouping reagents. We then found that a new batch of reagents was giving reactions that were not clear to read. This prompted investigation of the effect of different variables on the method.

Table IV. Strains used in investigations of Phadebact® grouping.

Strain	NCTC reference	Lancefield group as given by Reference Laboratory	
1	8198	A	S. pyogenes
2	2218	A	S. pyogenes
3	8370	A	S. pyogenes
4	8181	B	S. agalactiae
5	6175	B	S. agalactiae
6	8100	B	S. agalactiae
7	4335	C	S. dysgalactiae
8	9682	C	S. equi
9	5371	C	S. equisimilis
10	8307	D	S. durans
11	775	D	S. faecalis
12	7171	D	S. faecium
13	5969	G	Streptococcus species
14	6198	G	Streptococcus species
15	7932	G	Streptococcus species

Table V. Broths used in investigations of Phadebact® grouping.

Broth	Nature and origin of broth
1	Southern Group nutrient broth
2	Southern Group glucose broth
3	BBL nutrient broth
4	BBL Todd-Hewitt broth
5	Oxoid nutrient broth
6	Oxoid Todd-Hewitt broth
7	Difco nutrient broth
8	Difco Todd-Hewitt broth
9	Gibco nutrient broth
10	Gibco Todd-Hewitt broth
11	Gibco glucose broth

Table VI. Characterization of results in investigations of Phadebact® grouping.

True reactions	+ + +	Very strong co-agglutination
	+ +	Strong co-agglutination
	+	Moderate co-agglutination
	±	Weak co-agglutination
	−	No reaction
False reactions	+ L	Lumpy reaction
	+ LA	Lumpy reaction with true co-agglutination
	+ c	Co-agglutination with the wrong reagent
	+ g	Granular reaction

For the investigation, 15 strains of β-haemolytic streptococci (three strains each of Lancefield groups A, B, C, D and G) were obtained from the National Collection of Type Cultures (NCTC) (Colindale, U.K.) (Table IV).

Eleven commercially available broth media were examined (Table V). The media were prepared strictly according to the manufacturers' instructions and were distributed in volumes of 2, 4, 6 and 10 ml. Each strain was inoculated into each volume of each broth (i.e., 15 strains × 11 broths × 4 different volumes) and incubated at 37 °C overnight. The pH value of each broth was recorded before inoculation and again after overnight growth. A surface viable count was also performed on one of the strains from the 10 ml volume of each broth after overnight growth under the test conditions.

Each of the 15 strains growing in each volume of each broth was grouped by the Phadebact® method, in strict accordance with the manufacturers' instructions. A representative of each group in each broth was also checked by Lancefield's method (acid extraction).

The effect of diluting the broth culture after growth was investigated. Doubling dilutions, extending to 1 in 16, of 6 and 10 ml volumes of each broth inoculated

Table VII. Phadebact® grouping, strain 5 (Lancefield group B).

Broth	Volume (ml)									
	2					10				
	A	B	C	D	G	A	B	C	D	G
1	−	+ + +	−		+ + g	−	+ + +	−		+ + g
2	−	+ + +	−		+ + g	−	+ + +	−		+ + g
3	−	+	−		−	−	+ +	−		−
4	−	±	−		± g	+ L	+ + +	+ L		+ + g
5	−	+ +	−		± g	−	+ +	−		± g
6	−	+ + +	−		+ g	−	+ + +	−		+ + g
7	−	+	−		−	−	+	−		−
8	−	+ + +	+ + c		± g	−	+ + +	+ + c		+ + g
9	−	+	−		−	−	+ +	−		−
10	−	+ + +	+ + c		− g					
11	−	+ + +	−		± g	−	+ + +	−		+ + g

Table VIII. Relationship between pH and viability counts for different broths and grouping by Phadebact® co-agglutination method.

Broth	Best Phadebact® results	Nature and origin of broth	Initial pH	Final pH	pH change	Organisms/ml strain 1
1	*	Southern Group nutrient broth	6.56	6.41	0.15	5.3×10^9
6		Oxoid Todd-Hewitt broth	7.67	6.26	1.41	2.0×10^8
11	*	Gibco glucose broth	7.07	6.42	0.65	1.9×10^8
2	*	Southern Group glucose broth	6.42	6.08	0.34	1.1×10^8
10		Gibco Todd-Hewitt broth	6.87	6.12	0.75	7.8×10^7
8		Difco Todd-Hewitt broth	7.95	7.15	0.80	7.4×10^7
4	*	BBL Todd-Hewitt broth	7.66	5.34	2.32	5.5×10^7
7		Difco nutrient broth	6.87	7.18	−0.31	3.0×10^7
5		Oxoid nutrient broth	7.03	6.90	0.13	2.3×10^7
3		BBL nutrient broth	6.92	7.09	−0.17	1.9×10^7
9		Gibco nutrient broth	6.83	7.34	−0.51	6.9×10^6

with each strain were made in phosphate buffer (pH 7.4) and each of these was grouped by the Phadebact® method.

Co-agglutination reactions were read ultracritically and every type of reaction visible was recorded. Reading was so meticulous because there had been criticism that, in inexperienced hands, there could be problems in distinguishing true co-agglutination reactions.

The scale for recording reactions is shown in Table VI. Among the false-positive reactions several patterns were seen. Very occasionally a true co-agglutination reaction was recorded with the wrong reagent (designated as +c). This constitutes a genuine false-positive reaction. Two other types of reaction were seen, lumpy (L) and granular (g), both of which were clearly distinct from the true co-agglutination reaction. Occasionally a lumpy reaction could be seen overlying a true co-agglutination reaction and this was designated LA. An example of the results recorded is shown in Table VII. The relationship between pH and surface viable count in different media and results of grouping using the Phadebact® kit are shown in Table VIII.

DISCUSSION

Although *Streptococcus pyogenes* (Lancefield group A) is universally recognized as a major pathogen, it has long been accepted that there is a need to group streptococci other than those of group A. For example, group B streptococci are involved in neonatal sepsis [10] and group C streptococci in acute and subacute endocarditis [11] while there is an association between streptococci of group G and pharyngitis [12]. The epidemiological importance of non-group A streptococci has recently been clearly shown by several workers [13–16].

However, the time-consuming and technically exacting procedures of the Lancefield precipitation method or Fuller's formamide modification [17] may discourage clinical laboratories from serological grouping of β-haemolytic streptococci. Hamilton and Martin [18] in a recent survey in the United Kingdom found that 7.4% of 107 laboratories used bacitracin discs alone to screen for group A and non-group A β-haemolytic streptococci.

A truly rapid reliable method may encourage the grouping of β-haemolytic streptococci, even in small hospitals. Since we had found the Phadebact® co-agglutination technique simple to perform and interpret we decided to make further studies rather than abandon the method when difficulties were experienced with a new batch of reagents.

Our studies with various broths showed that two factors appear to be involved in relation to pH: its degree of change during incubation and its final value at the time of grouping. Broth 4, which we found best, had a high initial pH (7.66) but showed a considerable fall to the most acid pH (5.34) after incubation. In contrast, broth 6, which had a very similar starting pH to broth 4, showed a smaller drop and, consequently, a less acid final pH (6.26). This broth gave much poorer grouping results. Broth 2 showed a small change in pH during incubation (0.34 units) but was more acid (pH 6.08) than 6 and gave good grouping results. Three broths (3, 7 and 9) rose in pH during incubation. It was with these broths that grouping failures occurred.

In broths where viable counts were less than 5×10^7 organisms/ml fewer successful groupings occurred. Three of these broths were those in which the pH rose on incubation while the fourth, broth 5, had a final pH of 6.9, only 0.13 units less than the initial pH.

Dilution of the culture after growth and immediately prior to grouping tended to diminish reactions of all types and is not recommended.

As a result of our studies we recommended broth 4, Brooke Bond Liebig (BBL) Todd-Hewitt broth, for routine use.

In our study, the Phadebact® kit grouped more isolates than our conventional Lancefield method, possibly due to its sensitivity and the comparative simplicity of the method. The Phadebact® method has three important advantages. Firstly, it is labour- and time-saving which is valuable in a busy routine laboratory. Secondly, if the test is performed on a four-hour culture, or directly from the primary culture plate, it is quick and the clinician can be informed of results within 24 hours of receipt of the specimen, making the system amenable to use at weekends and public holidays. Thirdly, we have found it possible to obtain reliable results from mixed cultures with two types of β-haemolytic streptococci, or with streptococci mixed with other organisms.

Because the Phadebact® kit does not contain a reagent for group D streptococci and confusion may be caused by group D strains cross-reacting with other antisera, testing for esculin hydrolysis, which can be determined four hours after inoculation, is strongly recommended. This can be done economically by dividing a bile-esculin agar plate into eight segments and inoculating it at the same time as the grouping broth.

ACKNOWLEDGEMENTS

We should like to thank Mr P. Wells, Mrs I. Diaz and Miss A. Bhamra for their immaculate technical assistance, Miss P. Townsend for her secretarial work and Pharmacia (Great Britain) Ltd. for providing the materials for this study.

REFERENCES

1. Lancefield, R.C. (1933): A serological differentiation of human and other groups of haemolytic streptococci. *J. exp. Med.*, *57*, 571.
2. Riskaer, N. (1943): Thesis, Copenhagen.
3. Rosendal, K. (1956): Grouping of hemolytic streptococci belonging to groups A, C, and G. *Acta. Pathol. Microbiol. Scand.*, *39*, 127.
4. Kronvall, G. (1973): A rapid slide-agglutination method for typing pneumococci by means of specific antibody adsorbed to protein A-containing staphylococci. *J. Med. Microbiol.*, *6*, 187.
5. Christensen, P., Kahlmeter, G., Jonsson, S. and Kronvall, G. (1973): New method for the serological grouping of streptococci with specific antibodies adsorbed to protein A-containing staphylococci. *Infect. Immun.*, *7*, 881.
6. Farrell, B. and Amirak, I.D. (1976): Agglutination grouping of streptococci. *Lancet*, *II*, 1082.
7. Rosner, R. (1977): Laboratory evaluation of a rapid four-hour serological grouping of groups A, B, C and G beta-streptococci by the Phadebact Streptococcus test. *J. Clin. Microbiol.*, *6*, 23.
8. Finch, R.G. and Phillips, I. (1977): Serological grouping of streptococci by a slide coagglutination method. *J. Clin. Pathol.*, *30*, 168.
9. Anthony, B.F., Perlman, L.V. and Wannamaker, L.W. (1967): Skin infections and acute nephritis in American Indian children. *Pediatrics*, *39*, 263.
10. Eickhoff, T.C., Klein, J.O., Daly, A.K., Ingall, D. and Finland, M. (1964): Neonatal sepsis and other infections due to group B streptococci. *N. Engl. J. Med.*, *271*, 1221.
11. Finnegan, P., Fitzgerald, M.X., Cumming, G. and Geddes, A.M. (1974): Lancefield group C streptococcal endocarditis. *Thorax*, *29*, 245.
12. Hill, H.R., Wilson, E., Caldwell, G.G., Hager, D. and Zimmerman, R.A. (1969): Epidemic of pharyngitis due to streptococci of Lancefield group G. *Lancet*, *II*, 371.
13. Feingold, D.S., Stagg, N.C. and Kunz, L.J. (1966): Extrarespiratory streptococcal infections: importance of various serological groups. *N. Engl. J. Med.*, *275*, 356.
14. Horstmeier, C. and Washington, J.A. (1973): Microbiological study of streptococcal bacteremia. *Appl. Microbiol.*, *26*, 589.
15. Howard, J.B. and McCraken, G.A., Jr (1974): The spectrum of group B streptococci infections in infancy. *Am. J. Dis. Child.*, *128*, 815.
16. Reinarz, J.A. and Sanford, J.P. (1965): Human infections caused by non group A or D streptococci. *Medicine (Baltimore)*, *44*, 81.
17. Fuller, A.T. (1938): The formamide method for the extraction of polysaccharides from haemolytic streptococci. *Br. J. Exp. Pathol.*, *19*, 130.
18. Hamilton, W.J. and Martin, R.J. (1975): A survey of clinical microbiological techniques used in the United Kingdom. *Med. Lab. Technol.*, *32*, 307.

Discussion

R. Facklam (Atlanta, U.S.A.): Does esculin refer to bile-esculin plates?

R. J. Williams: Yes.

R. Facklam: Secondly, while doing the study, was there ever one strain that was grown in all the broths at the same time, so that when the co-agglutination tests were done the operator's reading of the reactions was influenced by knowing the identity of the strain?

R. J. Williams: It was difficult to keep it absolutely blind but we tried to randomize as far as possible and both the strains and the broths were always referred to only by code numbers. Although the code was not broken until the end of the study, so many groupings were performed that after a while operators began to recognize the identity of the strains.

R. Facklam: Do you feel that an unbiased judgement was made about each individual reaction as far as possible?

R. J. Williams: Yes, the judgement was as unbiased as possible.

R. Facklam: I have tested trypticase-soy broth and it works very well with the co-agglutination reagent. This would be another, different broth which could be used and which does not give the broth cross-reactions that are seen with other types of broth.

H. Arvilommi (Jyväskylä, Finland): Why were volumes as large as 10 ml used?

R. J. Williams: We wanted to find out whether growing the strains in different volumes would make any difference to the results. In fact, we found it was better to use smaller volumes.

J. Rotta (Prague, Czechoslovakia): First, I was rather surprised by the large number of non-groupable strains using the HCl extraction technique. There were about 50 strains, of which 13 were non-groupable by HCl. Is there an explanation? It is rather difficult to understand. Is it because of the poor sera that have to be used, which were commercial, were they not?

R. J. Williams: Yes.

J. Rotta: Secondly, and this relates to a number of other papers as well as that of Dr Williams, more attention should be paid to the strain which is used for raising the sera. Suppose we have group A strain which is used for immunization studies, if it contains the T-antigens, such as 2, 4 or 8, which may also be present in groups C and G, there will of course be cross-reaction by the other groups as well.

R. J. Williams: If we were able to raise our own antisera and perform a co-agglutination method on that, these problems could be avoided. Clearly, this is not practical in routine clinical laboratories.

W. R. Maxted (London, England): It is refreshing to see this problem approached so sensibly. Was the collection of group D strains studied all haemolytic?

R. J. Williams: I cannot remember offhand, but I do not think that they were.

W.R. Maxted: If they were not haemolytic, they would not have been a worry in this context.

R. J. Williams: No.

I. D. Amirak: We wanted to look at group D strains to examine the problem of cross-reactions. Consequently, we went to each bench at the end of the day and collected routinely isolated strains. These were simply 50 consecutive group D strains picked up regardless of their haemolytic properties.

W. R. Maxted: I see. Thank you. How was the Lancefield test done? Was it a layering technique in capillaries?

R. J. Williams: It was a layering technique.

I. D. Amirak: Perhaps I might try to answer Professor Rotta's question as to why we could not group so many strains with the Lancefield technique. This could reflect two things, firstly, the quality of the antiserum, which is obtained from Wellcome and stored in the refrigerator, and secondly, although I am sorry to have to say it, sounding critical of my own laboratory in which we try to keep standards high, when there are a lot of junior technical staff and about 180,000 specimens are being processed each year there is no way in which a senior person can be available merely to supervise a delicate and exacting method such as Lancefield's method involving layering techniques in capillaries.

When the junior technical staff report a strain as non-groupable, unless there are strong reasons to doubt the result, senior staff do not ask for it to be repeated. Indeed, it is very difficult and time-consuming to re-check every doubtful or non-groupable answer in a busy routine laboratory. With Phadebact® it took one or two minutes to perform the test and if there was any doubt, technical or otherwise, it could easily and simply be repeated. I am sure any large clinical laboratory which performs a large number of tests has similar problems. For example, if there are 30 Lancefield groupings to perform between 10 a.m. and 3 p.m. in any day, unless the method is really simple and rapid, misreading of results will slip by. I cannot therefore blame the reagents alone. I think it is basically a problem of technical complexities which will exist in any large laboratory.

A procedure for identification of streptococci in clinical laboratory routine (application to 600 strains)

M. Pons and Y. Peloux

Hospital of the Conception, Bacteriological Laboratory, Marseilles, France

The importance of specific identification of streptococci in clinical routine is the reason why every possible means is sought to perform it with minimum effort and in the least possible time. Our method is as follows.

Any sample likely to contain streptococci is plated on Columbia CNA agar (colistin, nalidixic acid) with 5% sheep's blood. If microscopic examination allows recognition of a monomorphic flora consisting of streptococci, bacitracin and optoquine discs are placed directly on the plate. In addition a streak is made with a β-haemolysin-producing staphylococcus to investigate the possibility of a CAMP-factor (Fig. 1). If the flora is mixed, these presumptive tests are delayed by 24 hours.

After obtaining a pure culture several possibilities arise:
a) If the streptococcus is β-haemolytic, we use a method (which we have described in a previous paper [1]) very similar to the technique proposed by Wautelet and co-workers [2]. A bacterial emulsion is placed for one hour at 40°C in a Lytase solution (BBL) or Maxted's solution, *Streptomyces albus* enzymes. From this extract we undertake reverse agglutination with Phadebact® Streptococcus Test reagents

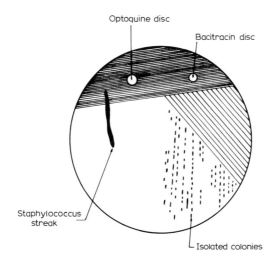

Optoquine disc

Bacitracin disc

Staphylococcus streak

Isolated colonies

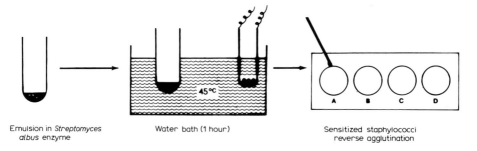

| Emulsion in *Streptomyces albus* enzyme | Water bath (1 hour) 45°C | Sensitized staphylococci reverse agglutination |

Fig. 2. Beta-haemolytic streptococci.

(Fig. 2). This method is simpler than the co-agglutination technique, the results are easier to read and it is also more rapid than reverse agglutination from a broth culture. A, B, C and G streptococci are identified in this way in a very short time (approximately one hour).

b) If the streptococcus is α- or non-haemolytic, *Streptococcus pneumoniae* is immediately identified owing to its susceptibility to optoquine and its identity is then confirmed by the bile solubility test.

As regards other streptococci, we use a range of media including:
– two 5% sucrose media, agar and broth, for investigation of glucan production,
– a 6.5% salt broth,
– a bile esculin agar,
– two cystine trypticase agar (CTA) media with, respectively, 1% mannitol and 1% sorbitol.

This small range of media is generally sufficient to identify enterococci and *S. salivarius* and to indicate the presumptive presence of *S. bovis, sanguis* or *mutans* species (Table I).

For dextran-producing strains and other streptococci not identified (either because they are not enterococci or because they are β-haemolytic streptococci not belonging to groups A, B, C or G) we use, in a second step, a more stringent range of tests including:
– testing for arginine hydrolysis (ADH) and acetoin production (VP),
– testing for acid production from a number of sugars.

Table I. Minimum range of media used in identification of streptococci.

Species	5% Sucrose	NaCl 6.5%	Bile-esculin	Mannitol	Sorbitol
S. faecalis	0	+	+	+	+
S. faecium	0	+	+	+	−
S. durans	0	+	+	−	−
S. salivarius	Levans	−	±	−	−
S. sanguis	Dextrans/−	−	−	−	−
S. bovis	Dextrans/−	−	+	−/+	−
S. mutans	Dextrans	−	−	+	+

Table II. Streptococcal species studied (origin and number).

Species	Stock strains	Hospital strains	Total
S. pyogenes	18	64	82
S. agalactiae	4	78	82
S. equisimilis	5	11	16
S. zooepidemicus	4	–	4
S. equi	3	3	6
S. dysgalactiae	2	–	2
S. faecalis	1	44	45
S. faecium, durans	6	15	21
S. bovis	2	25	27
S. equinus	2	–	2
S. avium	1	4	5
S. lentus	2	–	2
S. uberis	1	–	1
Species of group G	3	12	15
S. anginosus	–	17	17
S. sanguis I	1	43	44
S. sanguis II	3	29	32
S. salivarius	2	42	44
S. mutans	1	3	4
Species of group L	2	–	2
Species of group R	1	–	1
Species of group S	1	–	1
S. mitis	1	65	66
S. pneumoniae	–	15	15
S. milleri	2	50	52
Total	68	520	588
Undetermined		18	18
Total		538	606

As previously noted [3], tests of these characteristics are easily made using micro-methods.

If necessary, grouping according to the Lancefield method is performed.

RESULTS

More than 600 streptococci have been studied using this procedure, of which 68 were stock strains and 538 came from various pathological sources. Their distribution is presented in Table II.

Two hundred and twenty-two strains were β-haemolytic. Ninety-six per cent could be rapidly identified after the action of *S. albus* enzyme followed by use of Phadebact® Streptococcus Test reagents. The remaining strains belonged to groups F(6), L(1), P(1). One β-haemolytic strain could not be identified even through Lancefield grouping. Seventeen α- or non-haemolytic strains also remained un-identified.

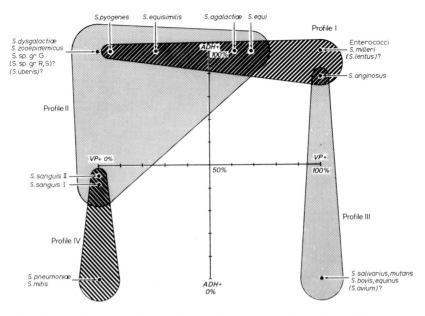

Fig. 3. Distribution of species with regard to profiles determined by ADH and VP characteristics.

Identification of the other strains caused no particular problem. We were able to identify several strains which are normally β-haemolytic but which on this occasion did not manifest this characteristic. The 'recovery' of normally β-haemolytic strains was made possible owing to their biochemical profile as presented in Fig. 3.

This figure results from the key significance we attach to the ADH and VP characteristics and enables four 'profiles' to be distinguished.

Profile I. Enterococci and *S. milleri* always present profile I. The species showing this profile are easily identified by their biochemical characteristics (Table III).

S. anginosus shows this profile in 88% of cases. When it belongs to group G it is identified by reverse agglutination.

S. equi, S. equisimilis and *S. agalactiae* often show this profile (60–70%). They can be identified by reverse agglutination. *S. lentus* was never encountered in our clinical material.

S. pyogenes is present in 4% of cases.

Profile II. In profile II, only α- or non-haemolytic strains raise problems. Dextran production then becomes a very important characteristic (Table IV) for differenti-

Table III. Identification of non β-haemolytic species presenting profile I (ADH + VP+).

Species	Haemolysis	Bile-esculin	NaCl 6.5%	Mannitol	Sorbitol
S. milleri	−(α)	−	−	−	−
S. faecalis	−(α)	+	+	+	+
S. faecium	α(−)	+	+	+	−
S. durans	α(−)	+	+	−	−

164

Table IV. Identification of species which may present profile II (ADH+ VP−).

Species	Haemo-lysis	5% Sucrose	Mann-itol	Sorb-itol	Esculin	Group antigen		
S. dysgalactiae	α	−	−	+/−	−	C		
Species of group R	γ	−	−	−	+	R		
S. suis	γ	−	−	−	+	S		
S. uberis	α	−	+	+	+	−		
S. sanguis I	α	+/−	−	−	+	H/−		
S. sanguis II	α	+/−	−	−	−	H/−		
						C:	Treh-alose	Lact-ose
S. zooepidemicus	β	−	−	+	−		−	+
S. equi	β	−	−	−	−		+	−
S. equisimilis	β	−	−	−	+/−		+	+/−

Table V. Identification of species presenting profile III (VP+ ADH−).

Species	Haemo-lysis	5% Sucrose	Mann-itol	Bile-esculin	Raffin-ose	Arabin-ose	Group antigen
S. salivarius	−	Levans	−	±	+	−	−
S. mutans	−	Dextrans	+	−	+	−	−
S. bovis	α/−	Dextr./−	−/+	+	+	−	D
S. equinus	α	−	−	+	−	−	D
S. avium	α	−	−	+(−)	−	+	D and Q

ating S. sanguis (S. sanguis II belonging to this profile in 47 % of cases and S. sanguis I in 43 % of cases) from the other α-haemolytic streptococci present in 100 % of cases in the profile but less frequently found in clinical routine (S. dysgalactiae, uberis, group R, group S). In most instances, β-haemolytic streptococci from A, C, G groups belong to this profile.

Profile III. Species belonging to profile III are generally identified by their biochemical characteristics. (However, in our study we had to test for possible presence of a group antigen) (Table V). They present this profile in 100 % of cases. In two cases S. anginosus group F was ADH-negative.

Profile IV. In profile IV, except for pneumococci, which are easily differentiated, hydrolysis of esculin, glucan production and, possibly, the presence of an H antigen enable specific differentiation (Table VI). S. mitis belongs to this profile in 100 % of cases.

In conclusion, it seems to us that our method is particularly inexpensive since only one blood agar plate is necessary for the identification of β-haemolytic streptococci. The grouping obtained with Phadebact® Streptococcus Test reagent is prompt and very reliable when preceded with a 60-minute extraction using Streptomyces albus enzymes.

Table VI. Identification of species which may present profile IV (ADH − VP−).

Species	Optoquine	Esculin	5% Sucrose	Group antigen
S. mitis	−	−	−	−
S. pneumoniae	+	±	−	−
S. sanguis I	−	+	Dextr./−	H or −
S. sanguis II	−	−	Dextr./−	H or −

One blood agar plate is likewise sufficient for the differentiation of pneumococcus via inhibition with an optoquine disc.

With six additional tubes, most of the species frequently isolated in human pathology are also identified: *S. faecalis, S. faecium* and *S. salivarius*. However, as regards other species (*S. milleri, S. mitis, S. sanguis,* to mention only the most common), differentiation requires investigation of a greater number of characteristics among which ADH and VP are those of greatest significance.

REFERENCES

1. Pons, M. and Peloux, Y. (1979): Le groupage des Streptocoques beta-hémolytiques par la méthode Phadebact: intérêt de l'extraction préalable par la Lytase. *Ann. Microbiol., 130A,* 281.
2. Wautelet, S., Wauters, G. and Bruynoghe, G. (1975): Détermination des streptocoques au moyen d'anticorps fixés à la 'Protéine A' staphylococcique. *Ann. Microbiol., 126 B,* 251.
3. Pons, M. and Peloux, Y. (1978): L'identification des Streptocoques en pratique hospitalière. *Feuill. Biol., 19,* 27.

The co-agglutination technique: principle, theory and practical application, with consideration of staphylococcal agglutination and protein A precipitation

G. Kronvall

Department of Medical Microbiology, University of Lund, Lund, Sweden

INTRODUCTION

Visualization of antigen-antibody reactions in serological work can be achieved by coating specific antibodies onto particles. For this purpose, protein A-containing *Staphylococcus aureus* organisms provide a suitable type of reagent particle [1]. Firstly, protein A on the staphylococcal cell surface will specifically bind large amounts of immunoglobulin G from the antiserum source with no prior purification of IgG required. This uptake is mediated by a non-immune reactivity for the Fc part of IgG [2, 3]. Secondly, absorbed antibodies will become oriented with their antigen-combining sites directed outwards with a resultant high binding efficacy for the

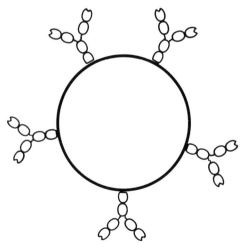

Fig. 1. Schematic drawing of a co-agglutination reagent particle consisting of stabilized S. aureus cell coated with IgG from rabbit antiserum through Fc binding to protein A. leaving the antigen binding Fab parts directed outwards.

167

corresponding antigen (Fig. 1). Antibody-coated staphylococci as a serological reagent were first described in a method for typing pneumococci [1]. The mixed clumping obtained gave rise to the term co-agglutination which is now generally used for the technique. A more general applicability of the co-agglutination method was recognized in the original description of the method, with preliminary positive results for viral particles and soluble antigens also. The co-agglutination technique has since been used for streptococcal grouping [5], serotyping of mycobacteria [6], identification of *Neisseria gonorrhoeae* [7], *N. meningitidis* [8], Salmonella, Shigella [9], *Pasteurella multocida* [10], T-typing of group A streptococci [11], typing of group B streptococci [12] and of *N. meningitidis* [8, 13]. Recently, the co-agglutination technique has been adopted for the detection of heat-labile *Escherichia coli* enterotoxin, a soluble bacterial product [14]. Several studies have been published on these applications with technical improvements allowing more rapid procedures and comparisons have been made with other established techniques [15–30].

Present applications of the co-agglutination method are largely limited to laboratory identification tests on isolated bacteria or their products. The direct demonstration of bacteria in pathological specimens has been tried using co-agglutination but the results have been relatively disappointing [31–35]. The main problem is a direct agglutination of the staphylococci by human serum, secretions or other human fluids or specimens. Prior treatment of the samples with soluble protein A will eliminate this direct agglutination but renders the technique complicated and expensive [31, 33, 34]. This indicates, however, that protein A on the staphylococcal reagent particles is involved in the direct clumping by human specimens. The present studies were aimed at the elucidation of mechanisms involved in such protein A-mediated clumping of staphylococci by human body fluids but not by rabbit sera used for making co-agglutination reagents. This discrepancy is even more intriguing because of the similar proportion of reactive IgG in human and rabbit serum.

MATERIALS AND METHODS

The cultivation and preparation of stabilized staphylococci for use in co-agglutination tests followed procedures described previously [1]. Myeloma serum samples with high levels of IgG-1 myeloma proteins were selected for the present studies. Subclass assignment was obtained through Gm-typing and ensured uniform Fc-mediated protein A reactivity [36]. Their capacity to precipitate soluble protein A was determined in gel diffusion experiments [36]. Similar results were obtained using heat-extracted protein A with a molecular weight of about 21,000 [37] or commercially available protein A (Pharmacia Fine Chemicals, Uppsala, Sweden) purified from meticillin-resistant strains [38]. Soluble protein A was radiolabelled using the Bolton-Hunter reagent [40]. Binding studies were performed as described previously [41].

RESULTS

Agglutination of S. aureus by normal human and rabbit sera

Agglutinin titres of serum samples from 20 healthy humans were determined using slide agglutination with live *S. aureus* Cowan I organisms harvested the same day.

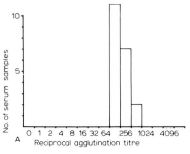

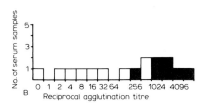

Fig. 2. A. Agglutination of S. aureus strain Cowan I by 20 normal human sera in slide tests. B. Agglutination of S. aureus strain Cowan I by 16 pathological sera containing elevated levels of IgG-1 myeloma proteins, all capable of Fc binding with protein A.

Eleven serum samples showed a reciprocal titre of 128, seven gave a value of 256 and two of 512. A histogram in Figure 2A shows this rather narrow range of titres of normal human sera. Parallel experiments with the almost protein A-negative S. aureus strain Wood 46 gave no agglutination in 15 cases. Three sera were positive, undiluted, and two showed reciprocal titres of 2 and 4, respectively. Nine serum samples from patients with staphylococcal infections were also tested with Cowan I organisms. Six osteomyelitis sera showed reciprocal titres of 64 (one case), 128 (two cases) and 256 (three cases), with no difference as compared to normals. Three serum samples from patients with furunculosis showed elevated titres with recip-rocal values of 4,096, 8,192 and 16,384, respectively. Corresponding reciprocal titres against Wood 46 were 4, 16 and zero.

In contrast to the titres obtained using human sera, rabbit serum samples were almost negative. Among five rabbit serum samples tested, the titres ranged between 1 and 8 in spite of a more than 90% reactivity of their IgG for staphylococcal protein A. Other protein A-carrying staphylococci were also tested for their suitability in the slide agglutination experiments. S. aureus strain Cowan III was grown and used live in parallel experiments with live Cowan I and also with formaldehyde- and heat-treated Cowan I organisms. No significant differences were noted in results from these comparative studies. All further experiments were therefore performed with stabilized Cowan I organisms unless otherwise stated.

Agglutination of S. aureus by human myeloma sera

Sixteen myeloma serum samples were selected on the basis of high levels of mono-clonal IgG of subclass 1 with low background polyclonal IgG. Both IgG alleles of the Caucasian types Gm (1) and Gm (4) were represented, as well as η and λ light chains. When tested in the slide agglutination method with S. aureus strain Cowan I and Cowan III, the titres obtained were not clustered at one level corresponding to the capacity of normal human IgG but showed a marked variation from no agglutina-tion at all, using undiluted serum, up to a reciprocal titre of 8,192 (Fig. 2B). All myeloma proteins were positively identified as being capable of binding protein A. Apparently, the simple view of protein A as an agglutinogen of staphylococci through its Fc binding capacity is erroneous [42]. Myeloma sera with as much as 57 g per litre of IgG capable of binding protein A might not agglutinate S. aureus organisms with a high density of protein A on their surface at all.

Correlation between staphylococcal agglutination and protein A precipitation

The myeloma sera studied were also analyzed regarding their mode of reactivity with soluble protein A in agar gel diffusion experiments, direct precipitation and the formation of soluble complexes seen as inhibition in the assay [43]. Seven myeloma proteins gave a precipitin reaction with protein A, whereas the other nine myeloma proteins formed soluble complexes. With staphylococcal agglutination titres ranging from 256 to 8,192 a striking correlation was noted with protein A-precipitating myeloma sera (Fig. 2B). In contrast, inhibiting myeloma proteins forming soluble complexes with protein A showed only low titres, ranging from zero to 512 (Fig. 2B). A similar correlation is seen with normal human and rabbit serum samples. Human sera, with reciprocal agglutination titres around 256, are capable of precipitating protein A, whereas rabbit serum samples, with titres around 4, do not precipitate protein A.

Agglutination of Cowan I by IgG fragments

Human gamma globulin was isolated from the serum of three apparently healthy individuals as well as from two normal serum pools and subjected to digestion with pepsin, producing $F(ab')_2$ fragments. The Cowan I agglutinin titres dropped to zero in all cases after pepsin digestion indicating the primary role of protein A-Fc interactions in the agglutination phenomenon studied.

Rabbit gamma globulin from an antiserum raised against the Cowan I strain was also pepsin-treated to produce $F(ab')_2$ fragments and then tested for agglutination of Cowan I organisms. The immune $F(ab')_2$ fragments were positive in these tests. Reduction and alkylation of the preparation to produce univalent Fab fragments resulted in complete loss of agglutinating activity. Controls using normal human gamma globulin were all negative.

Effects of carbamylation, hydrogen concentration and ionic strength on staphylococcal agglutination

In previous studies, chemical modification of immunoglobulin preparations by carbamylation was found to convert a protein A-precipitating capacity of a myeloma protein to an inhibiting one, with formation of soluble complexes [44]. Since protein A precipitation and staphylococcal agglutination seem to depend on similar mechanisms, the effect of carbamylation on agglutination was also studied. A pool of normal human serum samples diluted 1 : 2 was treated with 0.2 M KOCN for 30 and 120 minutes and 24 hours at pH 8.0 and 37 °C. An eightfold decrease in reciprocal agglutination titre was noted in parallel with a change from a precipitin reaction against soluble protein A to one of inhibition. Control experiments using 0.2 M KCl showed no changes due to the incubation alone. Carbamylation also did not alter the primary reactivity for protein A.

The effect of varying pH and ionic strength on staphylococcal agglutination by normal human serum was also studied. Buffers ranging from pH 4.0 to 9.0 were used for suspending the bacteria and for diluting normal human serum in the tests. At low pH values, agglutination was stronger with resultant higher titres. At pH 7.5 and above, lower titres were recorded. Differences were also noted for buffers of

170

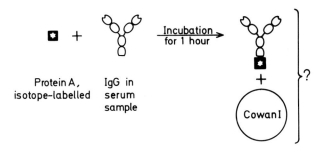

Fig. 3. Assay for secondary protein A-binding measuring binding of protein A-IgG complexes to protein A-containing S. aureus strain Cowan I. After addition of stabilized Cowan I particles, the bacteria were spun down and the radioactivity of the pellet was measured.

different ionic strength. At pH 5.0 and 8.0 a dilution of the buffers to 1:32 decreased the agglutination titres 32-fold and fourfold, respectively.

Staphylococcal binding of protein A-IgG complexes

A test method was devised to investigate the possibility of alternative protein A reactive sites on IgG as reported by others [45, 46]. Radiolabelled protein A was mixed with a serum aliquot and incubated at room temperature for one hour (Fig. 3). A 2 ml volume of 1 % stabilized, protein A-containing staphylococci was then added and the mixture centrifuged at 2,000 g. Bacteria-bound protein A was measured in a gamma counter and expressed as per cent of total radioactivity added. When normal human serum samples were tested at several dilutions, remarkably similar binding curves were obtained (Fig. 4A). With an excess of IgG over radiolabelled protein A there was a high uptake of radiolabelled protein A. Myeloma sera containing protein A-reactive IgG-1 myeloma proteins were also analyzed using the same technique. Myeloma sera capable of precipitating protein A were distinctly different from sera giving rise to soluble complexes. Precipitating myeloma sera showed higher binding levels of radiolabelled protein A (Fig. 4B). In contrast, all inhibiting myeloma sera gave very low binding levels as compared to normal human sera (Fig. 4C). This difference is even more pronounced when the elevated concentrations of IgG in myeloma sera are taken into account.

The effect of chemical modification of IgG was also tested using carbamylated normal human serum prepared as described above. Upon carbamylation of IgG the binding of protein A-IgG complexes was reduced in proportion to the degree of chemical modification (Fig. 5A).

The correlation between binding of protein A-IgG myeloma complexes to Cowan I organisms and the reciprocal agglutination titres of corresponding myeloma sera was also analyzed. As shown in Figure 5B, there was a positive correlation between the two parameters. In addition, a marked clustering of protein A-precipitating myeloma sera was noted separate from the inhibiting myeloma sera.

Further characterization of the observed binding of protein A-IgG complexes to protein A-containing staphylococci was obtained in another series of experiments. The influence of incubation time of radiolabelled protein A with IgG was measured using two protein A-precipitating IgG-1 myeloma sera. Binding levels

171

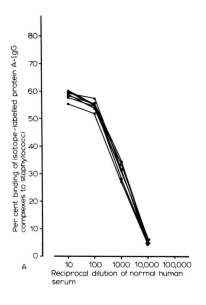

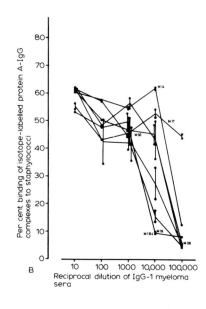

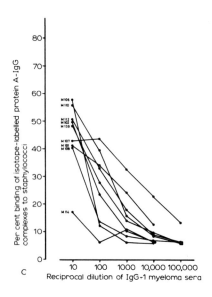

Fig. 4. A. Binding of radiolabelled protein A in complex with IgG of normal human sera to protein A-containing Cowan I organisms. All seven human serum samples tested show similar binding curves. B. Binding of radiolabelled protein A to S. aureus Cowan I after incubation with dilutions of myeloma sera containing IgG-1 paraproteins capable of precipitating protein A. C. Binding of radiolabelled protein A to S. aureus Cowan I, when complexes with IgG-1 myeloma proteins not capable of precipitating protein A.

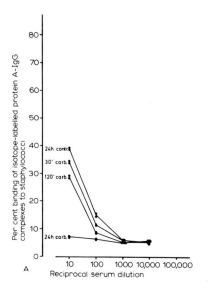
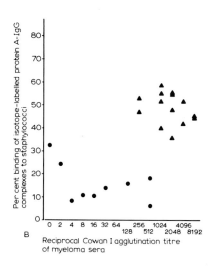

Fig. 5. A. The effect of carbamylation of normal human serum on the binding of radiolabelled protein A in complex with treated serum IgG to S. aureus Cowan I organisms. B. Correlation between per cent binding of radiolabelled protein A-IgG-1 myeloma protein complexes to S. aureus and the reciprocal Cowan I slide agglutination titres of the myeloma sera used. Protein A-precipitating myeloma proteins are marked with triangles and inhibiting proteins with closed circles.

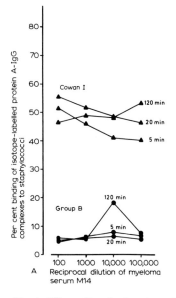
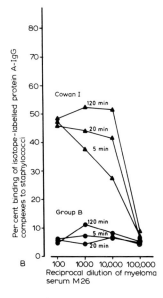

Fig. 6. Effect of incubation time of radiolabelled protein A with IgG-1 myeloma serum M14 (A) or M26 (B) on the subsequent binding of complexes to Cowan I or group B streptococcal organisms.

using myeloma serum M14 were affected differently at the two extremes of dilutions, 1:100 and 1:100,000 (Fig. 6A). The uptake values with excess protein A increased with incubation time. This was also reflected in controls using an Fc-negative group B streptococcal strain with measurable precipitation only at equivalence (1:10,000) and only after 120 minutes of incubation (Fig. 6A). The other protein A-precipitating myeloma serum showed an even more pronounced effect of incubation time on the binding of protein A-IgG complexes at low IgG levels, with no apparent differences with excess IgG (Fig. 6B). The results suggest that more than one mechanism might be operating giving rise to, or exposing, protein A-reactive structures on IgG other than the regular Fc-site.

DISCUSSION

The co-agglutination technique, as described originally for the serological typing of pneumococci [1], has been used for a wide range of practical serological applications mainly involving bacterial antigens [5–35]. In all different situations where the method is being used some basic requirements regarding the reagent materials have to be met in order to obtain successful results. Firstly, the antiserum used must be properly absorbed in order to produce clear-cut results without interfering cross-reactions. A co-agglutination reagent is only as good as the antiserum used for antibody coating. If an intracellular antigen caused problems in tests using extracts of bacterial cells, this cross-reactivity might not show up in co-agglutination using whole cells. Occasional exceptions of this kind nevertheless emphasize the common rule that the antiserum source used in the co-agglutination test must be adequately absorbed. From experience, this almost self-evident fact still has to be stated.

The second requirement relates to the fact that stabilized staphylococci used for making the co-agglutination reagents will absorb not only the specific antibodies but also other protein A-reactive immunoglobulins present. Since a minimum number of antigen-reactive IgG molecules has to be present on the particles in order to give rise to an agglutination reaction, the proportion of antigen-specific antibodies present in the antiserum must exceed this corresponding limit as determined by the total number of protein A molecules on the cells. Low-titred antisera might not work at all, therefore. High-titred antisera can often be diluted before coating the particles.

The third factor determining the quality of the co-agglutination reagent is the species origin of the antiserum used. In all present applications of the method, the antisera have been raised in rabbits. Coating staphylococci with rabbit IgG will not in itself cause a clumping of the bacteria, whereas a similar degree of coating using, for instance, human IgG, will give rise directly to agglutination. Differences between species in this regard depend on factors determining staphylococcal agglutination as well as protein A precipitation, two phenomena involving protein A which have been further analyzed in the present studies. Certain criteria can therefore be set up for the selection of experimental animals suitable for the production of antisera to be used in co-agglutination.

1) The staphylococcal agglutination titre of serum should not exceed 1:8.

2) Animal serum should not be capable of precipitating soluble protein A in agar gel diffusion experiments.

174

3) The majority of serum IgG or corresponding immunoglobulin class should be capable of combining with staphylococcal protein A through Fc structures [47, 48]. Based on these criteria, the rabbit and the mouse would fulfil the requirements for experimental animals. Other species such as the dog, pig, horse, rat and sheep are less suitable for antibody production.

Precipitation of soluble protein A has previously been shown to depend not only on the primary binding of protein A to the Fc part of immunoglobulins but also on other molecular interactions [44, 51]. Other studies support the concept that the primary reaction alone does not lead to precipitation. The molar composition of rabbit IgG-protein A complexes has been defined as IgG_2-pA_1, irrespective of the ratio of the two components [52, 53]. IgG is therefore functionally univalent in its Fc-reaction with protein A. Secondary interactions are directly suggested by the fact that presumably monovalent fragments of protein A precipitate with normal human IgG [54].

A three-component gel precipitation reaction, the star-precipitation, provides further insight into the mechanisms involved in protein A precipitation. In order to participate in this reaction, $F(ab')_2$ fragments have to be derived from IgG of a kind capable of precipitating protein A [56]. Apparently, Fab-located structures are involved in protein A precipitation [51, 55, 56]. The nature of this reactivity has been explained on the basis of experimental immunizations of animals as being anti-gamma globulin [57]. Recent demonstrations of normally occurring anti-gamma globulins in human sera might be taken as an indication that similar activities exist also in non-immunized individuals [58, 59]. However, the successful use of $F(ab')_2$ fragments from protein A-precipitating myeloma proteins in the star-phenomenon makes it less likely that some kind of anti-gamma globulin activity is involved [56]. Another explanation has recently been suggested by Endresen and co-workers [45, 60–62]. They demonstrated the existence of Fab-located, non-immune reactivity of low avidity with protein A [45, 60]. Such binding capacity has also been found in Fab fragments of pig immunoglobulin preparations [46]. Protein A-Fab reactivity was found to be necessary for protein A precipitation [62]. By analogy with mechanisms involved in protein A precipitation, we might therefore define two different types of interactions possibly involved in staphylococcal ag-glutination, one a participation of Fab-located protein A-reactive sites, the other involving anti-gamma globulin factors or protein-protein interactions of low avidity.

In the present investigations, we have confirmed that agglutination of *S. aureus* is mediated by protein A-immunoglobulin Fc interactions [42, 50]. This is demon-strated both by using different strains with and without protein A and by testing enzyme-produced immunoglobulin fragments. Results using myeloma sera indicate, however, that IgG Fc binding alone is not sufficient to produce clumping of the bacterial cells. For instance, a protein A-reactive IgG-1 myeloma protein with a serum concentration of 57 g per litre was completely incapable of causing any clumping of Cowan I (Fig. 7). Also, the results of others, where hyper-immune rabbit anti-*S. aureus* antiserum showed a 1,000-fold higher agglutinating capacity as compared to normal rabbit serum, in spite of the presence of approximately equal concentrations of protein A-reactive IgG, demonstrate, contrary to the con-clusions drawn, that protein A in itself is an incomplete agglutinogen [42]. Such a view is in agreement with other, earlier reports [49, 50]. It may be concluded that

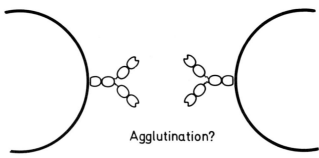

Agglutination?

Fig. 7. Schematic drawing of protein A-containing S. aureus organisms coated with IgG-1 myeloma proteins. In spite of identical constant heavy chain sequences with identical modes of Fc-mediated protein A-binding, the staphylococcal agglutination titres of a panel of IgG-1 myeloma proteins range from zero to 1:8192.

protein A-binding of IgG is a prerequisite for the agglutination of staphylococci but not in itself sufficient for the expression of the phenomenon. For further analysis of the mechanism of agglutination, we must study another phenomenon involving protein A, precipitation of soluble protein A in agar gel diffusion experiments. In the present studies, there was a direct correlation between tests for these two phenomena using normal human and rabbit sera as well as individual myeloma sera. In addition, chemical modification of IgG previously known to change the protein A-precipitating capacity [44] had a corresponding effect on agglutination of Cowan I organisms. Results regarding protein A precipitation might therefore also give some clues to the molecular events taking place in staphylococcal agglutination.

The existence of molecular events other than Fc binding operating in protein A-IgG interactions was tested using a different type of experimental arrangement (Fig. 3). Radiolabelled protein A was incubated with dilutions of serum samples and then mixed with *S. aureus* Cowan I organisms for possible binding to cell-bound protein A. When IgG was added in a 100-fold excess, where the primary complexes are of the IgG_2pA_1-type with no formation of spontaneous precipitate, a marked uptake of radiolabelled protein A was noted (serum dilution 1:10, Fig. 4A). Further studies using different myeloma sera and carbamylated IgG demonstrated that these experimental results also reflected the mechanisms underlying the agglutination and precipitation phenomenon. Since a high frequency of anti-gamma globulin activity cannot be expected in a panel of myeloma proteins, the binding of soluble protein A-IgG complexes to staphylococci is most likely mediated by Fab-located reactivity. Studies of the complex formation kinetics with excess IgG support such involvement of Fab binding structures. However, whereas the levels of binding of protein A complexes formed with excess IgG are not influenced by time of incubation, the uptake of complexes formed with precipitating myeloma proteins with excess protein A is highly time-dependent. Two mechanisms seem to be operating in some cases. The test system devised therefore indicates a higher degree of complexity in reactions involving protein A. The time-dependent secondary changes might correspond to protein-protein interactions or anti-gamma globulin activity [56–59].

The co-agglutination method has been used primarily for the identification of specific antigens of bacteria cultured in vitro or their products. Its potential application for the direct demonstration of pathogens in various clinical specimens has been realized but such a use is presently hampered by the direct agglutination seen with reagent particles [31–35]. Since several animal species show low levels of protein A-reactivity with no direct precipitation of protein A [47, 48], veterinary medicine ought to be one area where the co-agglutination method might be used for rapid clinical diagnostic applications.

The use of the co-agglutination method for rapid diagnosis of human pathogens in clinical specimens may also be foreseen. It requires a bacterial organism with intact Fc reactivity but no binding activity for the Fab part of IgG. Since myeloma proteins with identical Fc-located protein A reactivity show markedly different Fab binding to staphylococcal protein A, it follows that the structures responsible on the protein A molecule must be different. The test system described in the present studies seems well suited for the screening of staphylococcal strains or their mutants, as well as of other bacterial species with similar Fc reactivity [63], for lack of Fab reactivity. Such strains might serve as particles for use in the co-agglutination method for the rapid bedside diagnosis of various infectious diseases.

REFERENCES

1. Kronvall, G. (1973): A rapid slide agglutination method for typing pneumococci by means of specific antibody adsorbed to protein A-containing staphylococci. *J. Med. Microbiol.*, 6, 187.
2. Forsgren, A. and Sjöquist, J. (1966): Protein A from *S. aureus*. I. Pseudoimmune reaction with human γ-globulin. *J. Immunol.*, 97, 822.
3. Kronvall, G. (1967): Ligand-binding sites for streptolysin 0 and staphylococcal protein A on different parts of the same myeloma globulin. *Acta Pathol. Microbiol. Scand.*, 69, 619.
4. Kronvall, G. and Frommel, D. (1970): Definition of staphylococcal protein A reactivity for human immunoglobulin G fragments. *Immunochemistry*, 7, 124.
5. Christensen, P., Kahlmeter, G., Jonsson, S. and Kronvall, G. (1973): New method for the serological grouping of streptococci with specific antibodies adsorbed to protein A-containing staphylococci. *Infect. Immun.*, 7, 881.
6. Juhlin, I. and Winblad, S. (1973): Serotyping of mycobacteria by a new technique using antibody globulin adsorbed to staphylococcal protein A. *Acta Pathol. Microbiol. Scand. Sect. B, 81,* 179.
7. Danielsson, D. and Kronvall, G. (1974): Slide agglutination method for the serological identification of *Neisseria gonorrhoeae* with anti-gonococcal antibodies adsorbed to protein A-containing staphylococci. *Appl. Microbiol.*, 27, 368.
8. Zimmerman, S.E. and Smith, J.W. (1978): Identification and grouping of *Neisseria meningitidis* directly on agar plates by coagglutination with specific antibody-coated protein A-containing staphylococci. *J. Clin. Microbiol.*, 7, 470.
9. Edwards, E.A. and Hilderbrand, R.L. (1976): Method for identifying *Salmonella* and *Shigella* directly from the primary isolation plate by coagglutination of protein A-containing staphylococci sensitized with specific antibody. *J. Clin. Microbiol.*, 3, 339.
10. Rimler, R.B. (1978): Coagglutination test for identification of *Pasteurella multocida* associated with hemorrhagic septicemia. *J. Clin. Microbiol.*, 8, 214.
11. Hung, R.H. (1976): Comparison of the agglutination and the co-agglutination techniques in T-typing of *Streptococcus pyogenes*. *Acta Pathol. Microbiol. Scand. Sect. B*, 84, 109.
12. Kirkegaard, M.K. and Field, C.R. (1977): Rapid slide coagglutination test for identifying and typing group B streptococci. *J. Clin. Microbiol.*, 6, 266.

13. Danielsson, D. and Olcén, P. (1979): Rapid serotyping of group A, B and C meningococci by rocket-line immunoelectrophoresis and co-agglutination. *J. Clin. Pathol., 32,* 136.

14. Brill, B.M., Wasilunskas, B.L. and Richardson, S.H. (1979): Adaptation of the staphylococcal coagglutination technique for detection of heat-labile enterotoxin of *Escherichia coli. J. Clin. Microbiol., 9,* 49.

15. Edwards, E.A. and Larson, G.L. (1974): New method of grouping beta-hemolytic streptococci directly on sheep blood agar plates by coagglutination of specifically sensitized protein A-containing staphylococci. *Appl. Microbiol., 28,* 972.

16. Wautelet, J., Wauters, G. and Bruynoghe, G. (1975): Determination du groupe des streptocoques au moyen d'anticorps fixés à la 'protéine A' staphylococcique. *Ann. Microbiol., 126 B,* 251.

17. Menck, H. (1976): Identification of *Neisseria gonorrhoeae* in cultures from tonsillo-pharyngeal specimens by means of a slide co-agglutination test (Phadebact® Gonococcus Test). *Acta Pathol. Microbiol. Scand. Sect. B, 84,* 139.

18. Arvilommi, H. (1976): Grouping of beta-haemolytic streptococci by using coagglutination, precipitation or bacitracin sensitivity. *Acta Pathol. Microbiol. Scand. Sect. B, 84,* 79.

19. Hahn, G. and Nyberg, I. (1976): Identification of streptococcal groups A, B, C, and G by slide co-agglutination of antibody-sensitized protein A-containing staphylococci. *J. Clin. Microbiol., 4,* 99.

20. Deschamps, C., Ortenberg, M. and Perol, Y. (1976): Identification immunologique du groupe des Streptocoques par une technique rapide d'agglutination sur lame. *Médecine et Maladies Infectieuses,* 466.

21. Hryniewicz, W., Heczko, P.B., Lütticken, R. and Wannamaker, L.W. (1976): Comparison of three methods for grouping streptococci. *J. Clin. Microbiol., 4,* 28.

22. Rosner, R. (1977): Laboratory evaluation of a rapid four-hour serological grouping of groups A, B, C, and G beta-streptococci by the Phadebact Streptococcus Test. *J. Clin. Microbiol., 6,* 23.

23. Arvilommi, H., Kurasmaa, O. and Nurkkala, A. (1978): Rapid identification of group A, B, C, and G beta-haemolytic streptococci by a modification of the co-agglutination technique. Comparison of results obtained by co-agglutination, fluorescent antibody test, counterimmunoelectrophoresis, and precipitin technique. *Acta Pathol. Microbiol. Scand. Sect. B, 86,* 107.

24. Slifkin, M., Engwall, C. and Pouchet, G.R. (1978): Direct-plate serological grouping of beta-hemolytic streptococci from primary isolation plates with the Phadebact Streptococcus Test. *J. Clin. Microbiol., 7,* 356.

25. Stoner, R.A. (1978): Bacitracin and co-agglutination for grouping of beta-hemolytic streptococci. *J. Clin. Microbiol., 7,* 463.

26. Svenungsson, B. and Lindberg, A.A. (1978): Identification of *Salmonella* bacteria by co-agglutination, using antibodies against synthetic disaccharide-protein antigens 02, 04 and 09, adsorbed to protein A-containing staphylococci. *Acta Pathol. Microbiol. Scand. Sect. B, 86,* 283.

27. Szilagyi, G., Mayer, E. and Eidelman, A.I. (1978): Rapid isolation and identification of group B streptococci from selective broth medium by slide co-agglutination test. *J. Clin. Microbiol., 8,* 410.

28. Leland, D.S., Lachapelle, R.C. and Wlodarski, F.M. (1978): Method for rapid detection of group B streptococci by coagglutination. *J. Clin. Microbiol., 7,* 323.

29. Carlson, J.R. and McCarthy, L.R. (1979): Modified coagglutination procedure for the serological grouping of streptococci. *J. Clin. Microbiol., 9,* 329.

30. Olcén, P., Danielsson, D. and Kjellander, J. (1978): The laboratory identification of pathogenic Neisseria with special regard to atypical strains. An evaluation of sugar degradation, immunofluorescence and co-agglutination tests. *Acta Pathol. Microbiol. Scand. Sect. B, 86,* 327.

31. Olcén, P., Danielsson, D. and Kjellander, J. (1975): The use of protein A-containing staphylococci sensitized with anti-meningococcal antibodies for grouping *Neisseria meningitidis* and demonstration of meningococcal antigen in cerebrospinal fluid. *Acta Pathol. Microbiol. Scand. Sect. B, 83*, 387.

32. Suksanong, M. and Dajani, A.S. (1977): Detection of *Haemophilus influenzae* type B antigens in body fluid, using specific antibody-coated staphylococci. *J. Clin. Microbiol., 5*, 81.

33. Olcén, P. (1978): Serological methods for rapid diagnosis of *Haemophilus influenzae, Neisseria meningitidis* and *Streptococcus pneumoniae* in cerebrospinal fluid: A comparison of co-agglutination, immunofluorescence and immunoelectroosmophoresis. *Scand. J. Infect. Dis., 10*, 283.

34. Dirks-Go, S.I.S. and Zanen, H.C. (1978): Latex agglutination, counterimmunoelectrophoresis, and protein A co-agglutination in diagnosis of bacterial meningitis. *J. Clin. Pathol., 31*, 1167.

35. Thirumoorthi, M.C. and Dajani, A.S. (1979): Comparison of staphylococcal coagglutination, latex agglutination and counterimmunoelectrophoresis for bacterial antigen detection. *J. Clin. Microbiol., 9*, 28.

36. Kronvall, G. and Williams, R.C., Jr. (1969): Differences in anti-protein A activity among IgG subgroups. *J. Immunol., 103*, 828.

37. Jensen, K. (1958): A normally occurring staphylococcus antibody in human serum. *Acta Pathol. Microbiol. Scand., 44*, 421.

38. Kronvall, G. (1973): Purification of staphylococcal protein A using immunosorbents. *Scand. J. Immunol., 2*, 31.

39. Cohen, J.O. (1965): Studies of group agglutinogens of *Staphylococcus aureus. Nature (London), 205*, 917.

40. Langone, J.J., Boyle, M.D.P. and Borsos, T. (1977): ^{125}I protein A, applications to the quantitative determination of fluid phase and cell-bound IgG. *J. Immunol. Methods, 18*, 281.

41. Kronvall, G., Myhre, E.B., Björck, L. and Berggård, I. (1978): Binding of aggregated human β_2-microglobulin to surface protein structure in groups A, C and G streptococci. *Infect. Immun., 22*, 136.

42. Forsgren, A. and Forsum, U. (1972): Agglutination of *Staphylococcus aureus* by rabbit sera. *Infect. Immun., 5*, 524.

43. Kronvall, G., Quie, P.G. and Williams, R.C., Jr. (1970): Quantitation of staphylococcal protein A: Determination of equilibrium constant and number of protein A residues on bacteria. *J. Immunol., 104*, 273.

44. Kronvall, G., Messner, R.P. and Williams, R.C., Jr. (1970): Immunochemical studies on the interaction between staphylococcal protein A and γG globulin. *J. Immunol., 105*, 1353.

45. Endresen, C. (1979): Isolation of enzymatically derived fragments of porcine IgG and an examination of their reactivity against staphylococcal protein A. *Acta Pathol. Microbiol. Scand. Sect. C, 87*, 177.

46. Milon, A., Houdayer, M. and Metzger, J.J. (1978): Interactions of porcine IgG and porcine lymphocytes with protein-A sepharose. *Dev. Comp. Immunol., 2*, 699.

47. Kronvall, G., Seal, U.S., Finstad, J. and Williams, R.C., Jr. (1970): Phylogenetic insight into evolution of mammalian Fc fragment of γG globulin using staphylococcal protein A. *J. Immunol., 104*, 140.

48. Kronvall, G., Seal, U.S., Svensson, S. and Williams, R.C., Jr. (1974): Phylogenetic aspects of staphylococcal protein A-reactive serum globulins in birds and mammals. *Acta Pathol. Microbiol. Scand. Sect. B, 82*, 12.

49. Live, I. and Ranu, R.S. (1968): Serological activity of protein A of *Staphylococcus aureus*: the precipitinogen as an antigen for determining antibodies by the passive hemagglutination test. *J. Bacteriol., 96*, 14.

50. Kronvall, G. (1970): Protein A – an 'incomplete' staphylococcal agglutinogen. *Acta Pathol. Microbiol. Scand. Sect. B, 78*, 127.

51. Kronvall, G. (1971): *Interactions between Staphylococcal Protein A and γ Globulins*. Thesis. Studentlitteratur, Lund.

52. Hällgren, R., Stålenheim, G. and Bill, A. (1977): Elimination of protein A-IgG complexes from the blood circulation in rabbits: role of spleen and liver. *Acta Pathol. Microbiol. Scand. Sect. C*, *85*, 435.

53. Mota, G., Ghetie, V. and Sjöquist, J. (1978): Characterization of the soluble complex formed by reacting rabbit IgG with protein A of *S. aureus. Immunochemistry*, *15*, 639.

54. Hjelm, H., Sjödahl, J. and Sjöquist, J. (1975): Immunologically active and structurally similar fragments of protein A from *Staphylococcus aureus. Eur. J. Biochem.*, *57*, 395.

55. Kronvall, G. and Williams, R.C., Jr. (1971): The star phenomenon, a three component immunoprecipitation involving protein A. *Immunochemistry*, *8*, 577.

56. Lind, I. and Mansa, B. (1974): Immunochemical study of interactions between staphylococcal protein A, rabbit antistaphylococcal sera, and selected sera from non-immunized animals. *Scand. J. Immunol.*, *3*, 147.

57. Lind, I. (1974): The formation of antibodies against hidden determinants of autologous IgG during immunization of rabbits with *Staphylococcus aureus. Scand. J. Immunol.*, *3*, 689.

58. Johansson, M.E. and Espmark, J.Å. (1978): Elimination of inter-species reactive anti-IgG antibodies by affinity chromatography. *J. Immunol. Methods*, *21*, 285.

59. Nardella, F.A. and Mannik, M. (1978): Nonimmunospecific protein-protein interactions of IgG: studies of the binding of IgG to IgG immunoadsorbents. *J. Immunol.*, *120*, 739.

60. Grov, A. and Endresen, C. (1976): An examination of the 'star-phenomenon', a three component immunoprecipitation involving staphylococcal protein A. *Acta Pathol. Microbiol. Scand. Sect. C*, *84*, 333.

61. Endresen, C. (1978): Protein A reactivity of whole rabbit IgG and of fragments of rabbit IgG. *Acta Pathol. Microbiol. Scand. Sect. C*, *86*, 211.

62. Endresen, C. (1979): The binding to protein A of immunoglobulin G and of Fab and Fc fragments. *Acta Pathol. Microbiol. Scand. Sect. C*, *87*, 185.

63. Kronvall, G. (1973): A surface component in group A, C, and G streptococci with non-immune reactivity for immunoglobulin G. *J. Immunol.*, *111*, 1401.

Discussion

M. Slifkin (Pittsburgh, U.S.A.): With this particular technique, is it possible to make these organisms more sensitive by exposing more protein A? Could these organisms be treated in such a way that more antibody is made available on the staphylococci?

G. Kronvall: Since agglutination depends on the presence of a certain limited number of antibody molecules with their specificity and since it is very easy to get above that level, I do not think that any increase in the amount of protein A on the surface would improve anything in that regard.

J. Rotta (Prague, Czechoslovakia): Was any attempt ever made to check the protein A-binding capacity of antisera from animal species other than rabbits, such as sheep or horse?

G. Kronvall: We have investigated a large number of species. The Artiodactylae and the Perissodactylae are fairly poor with regard to their protein A-binding capacity. Among other orders there are occasional species with poor binding. For instance, the opossum-type animals are somewhat poor. There is only one bird species that will give any reactivity with protein A. This is the South American ostrich, and it shows an interesting reaction because it differs from ordinary Fc reactivity in specificity, but still shows the same criteria and phenomena as with mammalian binding to staphylococci.

Rapid identification of group A streptococci from the throat with selective enrichment broth and the Phadebact® Streptococcus Test

M. Slifkin and G. R. Pouchet-Melvin

Microbiology Section, Department of Laboratory Medicine, Allegheny General Hospital, Pittsburgh, Pennsylvania, U.S.A.

INTRODUCTION

The application of enrichment and selective media is advocated by many investigators as a means of obtaining efficient recovery of group A streptococci after subsequent subculture to blood agar medium. These media were also shown to be very effective when employed in conjunction with immunofluorescence for the rapid identification of β-streptococci from respiratory specimens [1–5]. Although the immunofluorescence technique for the identification of group A streptococci is considered to be an accurate and reliable method [3], the initial expense of the fluorescence microscopy equipment required may be costly for the small hospital laboratory [6].

The co-agglutination method has been shown to afford the clinical microbiologist a rapid and reliable approach to serogrouping β-haemolytic streptococci using direct plate and four-hour and 24-hour reading procedures [7–9], without the subjective interpretation associated with fluorescence microscopy. The combined use of a selective broth, Streptosel®, and Phadebact® co-agglutination test was recently reported to enable the detection and serogrouping of group A streptococci from throat specimens after a seven-hour incubation period [10].

We report here the development of a modified Todd-Hewitt broth that provides a relatively rapid means of determining the presence of group A streptococci in throat cultures with the co-agglutination method after five hours' incubation. The sensitivity of this medium-co-agglutination method for identification of group A streptococci from throat cultures was compared with direct plating, using duplicate swabs from individual patients.

MATERIALS AND METHODS

Stock cultures

Authenticated strains of group A streptococci were used for the initial evaluation of the medium. They comprised various strains of *Streptococcus pyogenes*, obtained from the American Type Culture Collection (ATCC). Various ATCC strains of viridans streptococci and stock cultures of various Enterobacteriaceae were also studied. All were maintained in trypticase soy broth containing 15 % (volume/volume) glycerol at −70 °C. These cultures were streaked on Columbia sheep blood agar plates to obtain isolated organisms and incubated at 35 °C.

Selective enrichment medium

Several basal media including Todd-Hewitt, brain heart infusion, heart infusion, infusion, trypticase soy and Bacto-tryptose broths, as well as Streptosel®, were first examined with regard to their relative growth efficiency for group A strepto-cocci. Group A growth-promoting agents, including Tween 80 [11], DL-cysteine HCl [12], dextrose and Bacto-peptone, were incorporated individually in Todd-Hewitt broth. Similarly, various other ingredients were incorporated separately in Todd-Hewitt broth in order to evaluate their respective inhibitory effects on a variety of organisms generally associated with normal throat flora as well as their inhibitory effects on the growth of group A streptococci. The inhibitors included sodium chloride [13], sodium azide [4], crystal violet [4], 8-hydroxyquinoline-5-sulfonic acid [14, 15], neomycin sulfate [16], and trimethoprim [17, 18]. Ferric citrate was incorporated in some instances in the modified Todd-Hewitt medium which contained the quinoline derivative as an inhibitor of viridans streptococci [19, 20].

Growth of stock bacterial strains

In order to examine the growth stimulation or inhibition of the various additives incorporated in Todd-Hewitt broth, several bacterial colonies were selected and inoculated into 10 ml of Todd-Hewitt broth until the optical density read between 1.0 and 1.5. The optical density was read at 650 nm in a spectrophotometer (Spectronic 100, Bausch and Lomb). Each bacterium was inoculated in triplicate into 25 ml of the experimental selective enrichment medium of modified Todd-Hewitt broth in 50 ml nephelometer flasks, so as to obtain an optical density reading between 0.05 and 0.10. Flasks containing control Todd-Hewitt broth were prepared in a similar manner. The flasks were incubated immobile or on a Rotary Orbital Shaker (Lab-Line Instruments, Inc.) adjusted to 100 revolutions per minute (rpm). All cultures were incubated at 37 °C and read at hourly intervals throughout six hours of incubation. The mean effective generation time was calculated and the lag time was estimated from the intercept of the exponential growth slope with the initial inoculation level [21].

Growth of clinical specimens

Pharyngeal cultures were obtained in duplicate by simultaneously rubbing two

Dacron-tipped swabs (Culturette II, Marion Scientific Corp., Kansas City, Mo.) over the throats of each of 527 patients. The swabs were sent to the laboratory.

One swab was rolled on a portion of a Columbia agar plate containing 5 % sheep blood (BBL). The inoculum was further distributed by streaking with a loop in order to obtain isolated colonies. The plates were then incubated at 35 °C under anaerobic conditions in Gas Pak (BBL) units for 12 to 18 hours. These plates were then observed for the presence of β-haemolytic streptococci by means of Gram staining and the determination of haemolytic and catalase activity.

The second of the paired swabs was inoculated into a 4 ml screw-capped vial containing 2 ml of a modified Todd-Hewitt broth. The vials were placed on an orbital shaker and rotated at 100 rpm for five hours at 37 °C. The vials were centrifuged at 3000 × g for 10 minutes. The supernatant was decanted and the pellet was delivered, using a Pasteur pipette, on a microscope slide. A drop of Phadebact® group A reagent was mixed with the pellet using a wooden applicator stick for a few seconds. The slide was then rocked for one minute and examined for co-agglutination using transillumination against a dark background.

Isolated colonies of β-haemolytic streptococci from each of the clinical primary plates and stock cultures of group A streptococci were serogrouped either by the direct [9] or four-hour co-agglutination procedures.

The identification of group A streptococci by conventional direct plating with serogrouping by co-agglutination was compared with identification using the five-hour modified Todd-Hewitt broth-co-agglutination method starting from the duplicate pharyngeal swabs from individual patients.

In another experiment, four throat swabs were obtained from patients with culture-established group A infections. Three of these swabs were inoculated into vials containing modified Todd-Hewitt broth, Todd-Hewitt broth or Streptosel® broth in order to compare the efficiency of each medium in yielding co-agglutination responses after five hours of incubation. The fourth swab was employed for the isolation of the group A streptococci on sheep blood agar plates.

The final formulation of the modified Todd-Hewitt broth (MTH) was next examined for its relative growth potential with a variety of bacteria as well as its clinical usefulness.

RESULTS

Various broths were prepared according to the manufacturers' instructions. Bacto-tryptose (Difco) was, however, prepared as a 5 % solution. The broths were examined in order to determine their respective growth potentials for group A streptococci throughout six hours of incubation and the basal medium to be employed throughout the investigation thus established. Todd-Hewitt broth proved the most satisfactory medium, yielding generation times significantly shorter than those observed with the other basal media (Table I). Further increases in the rate of growth of five strains of group A streptococci were observed when the cultures grown in Todd-Hewitt broth were shaken, as compared with stationary cultures (Table II). In an effort to stimulate the growth of various strains of group A streptococci in the basal medium further, alteration of the standard autoclave conditions was next examined. The use of 116 °C instead of 121 °C, combined with shaking of the

Table I. Growth rates [a] of group A streptococci in various culture broths.

Broth	Mean effective generation time (minutes)
Todd-Hewitt (BBL)	40.0
Tryptose (Difco)	86.0
Eugon (BBL)	106.6
Streptosel® (BBL)	166.6
Tryptose phosphate (Difco)	189.0
Lombard-Dowell with dextrose	213.3
Trypticase Soy (BBL)	259.9
Infusion (BBL)	308.1
Heart infusion (Difco)	308.6
Lombard-Dowell without dextrose	396.1
Bacto-Tryptose (Difco)	924.2

[a] Based on the means of at least three experiments with ATCC strain 2647; determined five hours post-inoculation; 100 rpm, 37 °C

cultures, yielded an observable shortening of the generation times of the group A streptococci examined (Table III).

Effect of various additives to the basal medium

Since the conditions for the promotion of relatively short generation times in the basal medium for the group A streptococci examined appeared to have been established, the incorporation, individually, in Todd-Hewitt broth of various streptococcal growth-stimulating agents was next examined with group A streptococci. When compared to the rate of growth in the basal medium without additives, stimulation of growth was particularly associated with Isovitalex, Tween 80 and Bacto-peptone. The addition of DL-cysteine HCl, dextrose and yeast extract did not effect marked stimulation of the group A streptococci within the six-hour period of observation. These data are summarized in Table IV.

Table II. Growth rates of various group A streptococci in stationary and shaken cultures with Todd-Hewitt broth.

Strain	Mean effective generation time [a] (minutes)	
	Stationary	Shaken [b]
21060	109.4	62.4
21546	122.3	51.3
21547	60.7	47.3
27263	118.8	69.1
Clinical isolate	67.1	50.1

[a] Based on the mean value of at least three experiments, determined five hours post-inoculation
[b] 100 rpm, 37 °C

Table III. Effect of autoclaving basal medium of Todd-Hewitt broth on growth of group A strepto-cocci.

Strain	Mean effective generation time [a] (minutes)	
	Autoclave temperature (°C)	
	116	121
ATCC 21546	93.5	244.6
ATCC 21547	94.5	173.3
ATCC 29263	116.3	181.7
Clinical isolate	120.2	195.3
Clinical isolate	189.0	271.3

[a] Mean values of at least three experiments determined five hours post-inoculation; shaken 100 rpm; 37 °C

In order to suppress the spectrum of bacteria generally thought of as normal flora, the incorporation, individually, of various bacterial inhibitors was next examined. Their effect on the growth rate of group A streptococci was also evaluated. Sodium chloride markedly suppressed the growth of the 10 strains of *S. aureus* used in this investigation. Suppression of growth of *S. uburis* and *Escherichia coli* by this agent was also observed. The addition of sodium azide reduced the generation time of *E. coli*. Although the quinoline derivative was responsible for a significant decrease in the growth rate of the viridans streptococci examined, it also slightly suppressed the growth rate of group A streptococci. The inhibitory effect of this agent on the growth of the viridans streptococci was, however, further increased by the incorporation of ferric citrate, without any further significant depression of growth of the group A streptococci. Crystal violet was extremely effective in lengthening the generation time of *S. aureus*. However, it slightly increased the generation time of group A streptococci. Neomycin and trimethoprim greatly

Table IV. Effect of various additives to Todd-Hewitt broth on increase in growth of group A streptococci.

Additive	Final concentration of additive (%)	Mean effective generation time (minutes) Strain of group A streptococci		
		ATCC 21546	ATCC 2647	ATCC 29263
None	–	54.7 [a]	77.7	54.0
Bacto-peptone	0.5	50.1	63.8	49.9
Cysteine HCl	0.02	112.4	73.9	81.3
Dextrose	0.1	100.1	110.7	120.3
IsoVitalex	1.0	46.8	72.7	49.5
Tween 80	0.05	47.3	57.7	41.7
Yeast Extract	0.5	120.0	53.1	57.1

[a] Mean values obtained from each of three experiments

186

Table V. Growth inhibition of bacteria with various additives in Todd-Hewitt broth.

Additive	Mean effective generation time [a] (minutes)			
	Group A streptococci (10)[b]	S. uburis (3)[b]	S. aureus (10)[b]	E. coli (10)[b]
None	54.3	61.30	61.6	51.2
Crystal violet	78.4	153.3	4,159.3	54.1
8-hydroxyquinoline-5-sulfonic acid	83.1	555.5	Nt[c]	Nt
8-hydroxyquinoline-5-sulfonic acid and 1 % ferric citrate	87.3	4,371.3	Nt	Nt
Sodium azide	60.7	67.3	65.5	1,013.3
Sodium chloride 1.4 %	70.5	167.5	3,501.2	181.3
Neomycin sulfate 0.002 %	225.2	Nt	Nt	Nt
Trimethoprim 0.01 %	195.3	Nt	Nt	Nt

[a] Means of five experiments per organism, determined five hours post-inoculation
[b] Number in parentheses refers to the number of strains examined
[c] Nt = not tested

suppressed the growth rate of group A streptococci. No growth of *Neisseria meningitidis* was detected, either in the basal Todd-Hewitt broth or with the various inhibitory agents. However, although *Haemophilus influenzae* did not grow in the presence of inhibitors, it grew slowly in Todd-Hewitt broth. These data are summarized in Table V.

The final formulation of MTH broth used in this investigation was based upon these preliminary findings (Tables I-V) and contained the enrichment and selective agents least inhibitory to group A streptococci.

All strains of group A streptococci examined exhibited some sensitivity to the inhibitory agents in MTH broth. Their generation times were approximately twice as long as those observed in standard Todd-Hewitt broth. In contrast, the generation times of these group A streptococci in Streptosel® broth were generally over an hour longer than those observed with MTH broth (Table VI). The members of the Enterobacteriaceae examined had generation times in either MTH or Streptosel® broth much longer than those in Todd-Hewitt broth, which are relatively short.

Table VI. Growth rates of various strains of group A streptococci in modified Todd-Hewitt broth, Todd-Hewitt broth and Streptosel ® broth.

Strain	Mean effective generation time (minutes)		
	Modified Todd-Hewitt	Todd-Hewitt	Streptosel®
ATCC 21546	82.1	39.9	159.9
ATCC 21547	106.9	40.7	169.8
ATCC 29263	71.5	49.8	132.1
Clinical isolate	39.2	52.6	277.3
Clinical isolate	102.9	43.3	148.5

*Table VII. Growth rates of various bacteria in modified Todd-Hewitt, Todd-Hewitt and Streptosel®
broths.*

Organism	Number of strains examined	Mean effective generation time (minutes)		
		Modified Todd-Hewitt	Todd-Hewitt	Streptosel®
Citrobacter diversus	5	1,386.3	44.7	1,387.7
Citrobacter freundii	5	1,188.3	40.7	1,188.3
Enterobacter cloacae	5	1,167.4	46.7	396.1
Escherichia coli	8	8,321.1	59.4	693.2
Haemophilus influenzae	5	No growth	1,039.9	4,160.5
Klebsiella pneumoniae	5	2,079.7	34.5	756.2
Neisseria meningitidis	5	No growth	No growth	No growth
Pseudomonas aeruginosa	5	4,170.8	1,386.3	4,460.5
Serratia liquefaciens	5	2,079.6	140.9	4,160.6
Staphylococcus aureus	5	1,386.3	61.6	2,772.6
Streptococcus pneumoniae	5	No growth	163.1	1,663.8

Furthermore, these generation times were much longer than those of group A
streptococci in either MTH or Streptosel® broths. MTH broth appeared more
inhibitory to *E. coli* and *Enterobacter cloacae* than Streptosel® broth. Both MTH
and Streptosel® broths suppressed the growth of *N. meningitidis*. The growth of
S. aureus and *H. influenzae* was also greatly affected by Streptosel® and MTH
broth. The growth of *H. influenzae* was completely suppressed by MTH broth during
the periods of observation. The growth of *Streptococcus pneumoniae* was completely
suppressed by MTH broth while growth of this bacterium in Streptosel®
broth was relatively slow (Table VII). MTH broth appeared to be more effective in

*Table VIII. Growth rates [a] of various viridans streptococci in modified Todd-Hewitt, Todd-Hewitt
and Streptosel® broths.*

Species	Modified Todd-Hewitt	Todd-Hewitt	Streptosel®
Streptococcus faecalis	85.8	39.6	Nt [b]
S. mitis, ATCC 15409	No growth	40.5	224.8
S. mutans, ATCC 25175	1,386.3	415.9	332.8
S. salivarius, ATCC 1341 9-2	1,188.9	102.7	220.7
S. salivarius, ATCC 27975	346.0	30.5	332.8
S. sanguis, ATCC 10557	No growth	24.4	141.0
S. sanguis, ATCC 10558	No growth	27.6	693.2
S. uburis, ATCC 9927	473.6	40.9	Nt
S. uburis, ATCC 13386	876.3	37.6	Nt
S. uburis, ATCC 13387	1,011.3	27.6	Nt
S. uburis, ATCC 19436	1,486.3	30.5	693.2

[a] Means of five experiments per organism, determined five hours post-inoculation
[b] Nt = not tested

Table IX. Comparison of co-agglutination results with three media inoculated with throat cultures containing group A streptococci.

Medium	Specimens [a]	Incubation time (hours)	Co-agglutination response Number positive	Number negative
Todd-Hewitt	20	5	0 (0 %)	20 (100 %)
Modified Todd-Hewitt	20	5	15 (75 %)	5 (25 %)
Streptosel®	20	5	2 (20 %)	18 (80 %)

[a] Four swabs per individual

lengthening the generation time of the viridans streptococci than Streptosel® broth. Three of the strains of viridans streptococci examined yielded no observable growth in MTH broth (Table VIII).

In a preliminary experiment prior to the clinical trials the efficiency of MTH broth was compated to that of Todd-Hewitt and Streptosel® broths using the throat swabs of 20 patients with culture-established group A streptococcal pharyngitis. Co-agglutination responses for group A streptococci were observed with 15 (75 %) of the patients when MTH broth was employed. In contrast, use of the other two media proved to be inadequate for the sero-identification of these bacteria when the co-agglutination method was applied (Table IX). It was also observed that the inoculated MTH broth vials were relatively less turbid with bacteria than the inoculated vials of Todd-Hewitt and Streptosel® broths.

A comparative evaluation of the MTH broth co-agglutination and standard culture co-agglutination methods for the sero-identification of group A streptococci from 527 throat swabs was next made. Agreement between the MTH co-agglutination (MTH-C) and direct culture co-agglutination methods occurred in 87 % of the specimens. However, 42.7 % of all the specimens were positive for group A streptococci by one method or both. The MTH-C method was negative when direct culture was positive in 21.7 %, while responses positive by MTH-C only occurred in 9.8 % of all positive specimens. These positive MTH-C responses were not false positives, as reculturing the throats of these patients yielded colonies of group A streptococci on blood agar plates.

The MTH-C method provided 80.4 % positive responses from the total positive and the standard culture co-agglutination method yielded 90.2 %. These results are summarized in Table X.

DISCUSSION

In order to formulate a broth medium that would effectively stimulate the growth of group A streptococci within a five-hour incubation period and suppress most of the normal bacterial flora, it was necessary to examine the various established culture, enrichment and selective broths. Furthermore, the various additives that have been previously reported as growth stimulants of group A streptococci as well as inhibitors of normal flora were also examined as to their relative efficacy. It became evident that Streptosel® broth was fairly effective in suppressing many normal flora bacteria, both Gram-negative and Gram-positive. The growth of group A streptococci in Streptosel® broth, as well as in Eugon broth was, how-

Table X. Comparative evaluation of modified Todd-Hewitt co-agglutination [a] and standard four-hour co-agglutination [b] procedures for detection of group A streptococci from throat cultures.

Standard 4-hour co-agglutination	Modified Todd-Hewitt-co-agglutination		Total
	+	−	
+	159	44	203
−	22	302	324
Total	181	346	527

[a] Five hours incubation; 37 °C, shaken 100 rpm; Phadebact® group A reagent
[b] 18-hour blood agar plate isolates; sero-identified with Todd-Hewitt four-hour Phadebact® procedure

ever, less rapid than in Todd-Hewitt broth. Furthermore, our data have shown that Streptosel® broth will not yield the minimum population of group A streptococci required for a co-agglutination response with Phadebact® group A reagent after a five-hour incubation period. This appears to be directly related to the poor growth of these group A streptococci in Eugon broth, the basic nutrient source of Streptosel® broth, as compared with Todd-Hewitt broth. Conversely, Todd-Hewitt broth alone was most effective in generating enough group A streptococci in a five-hour period for a visible co-agglutination response with Phadebact® group A reagent. However, our investigation has clearly indicated that neither Todd-Hewitt nor Streptosel® is as effective as modified Todd-Hewitt broth for obtaining a serodiagnosis of group A streptococci from a clinical throat specimen by means of co-agglutination. In connection with this observation, other investigators have considered that various common throat inhabitants, particularly the viridans streptococci, can interfere with the growth of group A streptococci [17, 18, 22, 23]. Although MTH broth does not yield generation times for group A streptococci as short as those observed with Todd-Hewitt broth, it does, however, provide a means of effective suppression of the many normal flora or 'background' organisms. The importance of suppression of these bacteria is especially significant to the co-agglutination response because some bacterial species, such as those of the coagulase-positive staphylococci group, can agglutinate with the normal globulins [24]. These bacteria could thus cause a nonspecific response with anti-group A streptococcal antisera and their growth should, therefore, be suppressed. It is also necessary to obtain enough group A streptococci in a broth milieu essentially devoid of 'background' organisms to yield an easily recognizable co-agglutination response when Phadebact® group A reagent is added.

Recovery of group A streptococci from agars and broth media is improved through the incorporation of selective agents, including sodium azide, crystal violet and antibiotics [17]. These agents very effectively eliminate the growth of 'diphtheroids', Neisseria, staphylococci and many Gram-negative rods but generally do not effectively inhibit the growth of the viridans streptococci [25].

In 1945 Pike developed a selective medium containing crystal violet and sodium azide for the pyogenic group of streptococci [4]. Crystal violet exhibits bactericidal activity towards Gram-positive bacteria, especially staphylococci. However, from our investigation, this agent appears somewhat inhibitory to group A streptococci and *S. uburis* within five hours of incubation. Sodium azide is bacteriostatic for Gram-negative bacteria but appears from this investigation to exert essentially no inhibitory effect on group A streptococci.

The addition of 8-hydroxyquinoline-5-sulfonic acid to broth media has been shown to suppress the growth of α-haemolytic streptococci [14, 15]. However, the incorporation of a ferric compound, such as ferric citrate, in MTH broth was required to potentiate further the inhibition by the quinoline derivative of the viridans streptococci. This requirement for various metallic ions of this chelating compound and its consequent enhanced antibacterial effect is well documented [19, 20].

The addition of sodium chloride to Todd-Hewitt broth was somewhat inhibitory to the growth of viridans streptococci. Its incorporation in conjunction with the quinoline derivative, however, resulted in a greater inhibition of the viridans strepto-cocci than with either of the two agents individually. Sodium chloride was also somewhat inhibitory to the growth of *S. aureus* in Todd-Hewitt broth. Previous investigations have shown that with concentrations of sodium chloride up to 12 % in broth there is some delay in but no failure of growth of *S. aureus* [26].

A 5 % sodium chloride-blood agar was previously reported to be more selective than sodium azide-crystal violet media for the isolation of group A streptococci [13]. In contrast, the results of our preliminary experiments have shown that a 5 % sodium chloride Todd-Hewitt broth significantly reduced the growth of all the strains of group A streptococci examined during a five-hour incubation period. The incorporation of 1.4 % sodium chloride with the quinoline derivative provided an effective combination which minimally suppressed the growth of group A streptococci while suppressing viridans streptococci and *S. aureus*.

A wide variety of antibiotics has been incorporated in agar media and enrichment broths as selective agents [27, 28]. The use of antibiotics in these media was shown to improve the recovery of group A streptococci by inhibiting certain members of the normal flora that occasionally overgrow and obscure the presence of β-haemolytic streptococci [29]. However, the viridans streptococci may not be effectively inhibited by antibiotic-containing media [15, 25]. Furthermore, certain antibiotics, such as neomycin [16], may suppress the growth of group A streptococci. The present investigation has demonstrated that neomycin or trimethoprim can significantly lengthen the generation time of the group A streptococci examined throughout a five-hour period of incubation. Thus, although antibiotics, as well as other selective agents, maximize the recovery of group A streptococci in selective agars after 18–25 hours of incubation, growth of group A streptococci in Todd-Hewitt broth is generally suppressed by these agents during a five-hour period of incubation. A formulation was, therefore, necessary which would decrease the toxicity of the various inhibitors to group A streptococci while maintaining a generation time for most group A streptococci that would provide enough organisms for a visible co-agglutination response after a five-hour incubation period. This was accomplished, in part, by the incorporation into MTH broth of various growth stimulants and inhibitors as well as by the use of shaken cultures and alteration of the autoclave temperature used in the preparation of the medium. Growth of *Haemophilus vaginalis* in a culture medium sterilized at 112 °C for 12 minutes has previously been shown to be improved in comparison with growth in medium sterilized at 121 °C for 15 minutes [30]. This observation, and ours, may be related to the lability of certain amino acids in the Todd-Hewitt broth base in the presence of the dextrose [32].

In contrast to other reports, the addition of yeast extract to Todd-Hewitt broth [31] did not result in a significant decrease during a five-hour period of incubation in the generation times of the majority of the group A streptococci strains examined. On the other hand, we have observed that the growth of these same bacterial strains increased significantly when the determinations were performed 12–18 hours after inoculation of MTH broth medium. It would appear, therefore, that the growth-potentiating factors associated with yeast extract are not effective until the late logarithmic phase of growth.

The MTH co-agglutination method (MTH-C) provides a rapid means of sero-identification of approximately 70 % of throat cultures containing group A strepto-cocci within a five-hour period of incubation. As in our laboratory, inoculation of the second of paired swabs onto a blood agar medium may be employed for the isolation and sero-identification of group A streptococci not detected by the MTH-C method. Furthermore, the direct plate co-agglutination method can be used for the isolation and identification of other β-haemolytic streptococci as well as for the examination of the total bacterial flora. The direct plate co-agglutination method may also be used to determine the relative amount of group A β-streptococci previously identified by the MTH-C method.

A report on group A streptococci issued five to six hours after culture has been obtained offers an advantage in the early diagnosis and treatment of streptococcal pharyngitis. This is unlikely to affect the duration or symptoms of pharyngitis or even the incidence of rheumatic fever. On the other hand, it is a convenience most physicians will consider useful, since it increases the likelihood of parents returning their children for proper therapy. Furthermore, this rapid identification procedure will be of great value in decreasing the workload often found within the clinical laboratory.

The MTH-C method can be used by technologists possessing varying bacterio-logical skills and can make it possible for a diagnostic laboratory to identify group A streptococci from the majority of throat cultures within the day on which the culture is collected.

REFERENCES

1. Facklam, R.R. (1976): A review of the microbiological techniques for the isolation and identification of streptococci. *CRC Critical Rev. Clin. Lab. Sci.*, 6, 287.
2. Facklam, R.R. (1978): *Isolation and Identification of Streptococci. Part I. Collection, Transport, and Determination of Hemolysis*, p. 3. Department of Health, Education and Welfare. Center for Disease Control, Atlanta, U.S.A.
3. Moody, M.D., Siegel, A.C., Pittman, B.P. and Winter, C.C. (1963): Fluorescent-antibody identification of group A streptococci from throat swabs. *Am. J. Publ. Health*, 53, 1083.
4. Pike, R.M. (1945): The isolation of hemolytic streptococci from throat swabs: experiments with sodium azide and crystal violet in enrichment broth. *Am. J. Hyg.*, 41, 211.
5. Redys, J.J., Parzick, A.B. and Borman, E.K. (1963): Detection of group A streptococci in throat cultures by immunofluorescence. *Public Health Rep.*, 78, 222.
6. Baron, E.J. and Gates, J.W. (1979): Primary plate identification of group A beta-hemolytic streptococci utilizing a two-disk technique. *J. Clin. Microbiol.*, 10, 80.
7. Lim, D.V., Smith, R.D. and Day, S. (1979): Evaluation of an improved rapid co-agglutination method for the serological grouping of beta hemolytic streptococci. *Can. J. Microbiol.*, 25, 40.

8. Rosner, R. (1977): Laboratory evaluation of a rapid four-hour serological grouping of groups A, B, C and G beta streptococci by the Phadebact Streptococcus Test. *J. Clin. Microbiol., 6,* 23.

9. Slifkin, M., Engwall, C. and Pouchet, G.R. (1978): Direct-plate serological grouping of beta-hemolytic streptococci from primary isolation plates with the Phadebact streptococcus test. *J. Clin. Microbiol., 7,* 356.

10. Slifkin, M. and Pouchet-Melvin, G.R. (1979): Rapid identification of group A streptococci from the throat with selective broth and the Phadebact Streptococcus test. *Annual Meeting of the American Society for Microbiology, Los Angeles,* Abstract No. C163.

11. Pine, L. and Reeves, M.W. (1973): Requirements of unsaturated fatty acids for aerobic growth of *Streptococcus pyogenes. Microbiol. Rev., 8,* 137.

12. Ginsburg, I. and Grossowicz, H. (1957): Group A hemolytic streptococci. I. A chemically defined medium for growth from small inocula. *Proc. Soc. Exp. Biol. Med., 96,* 108.

13. Mastromatteo, L. and Baldini, I. (1963): Simple medium for rapid screening and isolation of group A streptococci from contaminated materials. *J. Bacteriol., 86,* 1131.

14. Nakamizo, Y. and Sato, M. (1972): New selective media for the isolation of *Streptococcus haemolyticus. Am. J. Clin. Pathol., 57,* 228.

15. Sato, M. (1972): A new selective enrichment broth for detecting beta-hemolytic streptococci in throat cultures: quinoline derivate and three per cent salt as an additional agent to Pike's inhibitors. *Jpn. J. Microbiol., 16,* 532.

16. Vincent, W.F., Gibbons, W.E. and Gaafar, H.A. (1971): Selective medium for the isolation of streptococci from clinical specimens. *Appl. Microbiol., 22,* 942.

17. Gunn, B.A., Ohashi, D.K., Gaydos, C.A. and Holt, E.S. (1977): Selective and enhanced recovery of group A and B streptococci from throat cultures with sheep blood agar containing sulfamethoxazole and trimethoprim. *J. Clin. Microbiol., 5,* 650.

18. Kurzynski, T.A. and Meise, C. (1979): Evaluation of sulfamethoxazole-trimethoprim blood agar plates for recovery of group A streptococci from throat cultures. *J. Clin. Microbiol., 9,* 189.

19. Albert, A., Gibson, M.I. and Rubbo, B.D. (1953): The influence of chemical constitution on antibacterial activity. Part VI. The bactericidal action of 8-hydroxy quinoline (oxine). *Br. J. Exp. Pathol., 34,* 119.

20. Greenberg, J., Turesky, S.S. and Warner, V.D. (1976): Effect of metal salts in prolonging antibacterial activity of teeth treated with p-hydroxyquinoline. *J. Periodontol., 47,* 664.

21. Coultas, M.K. and Hutchison, D.I. (1962): Metabolism of resistant mutants of *Streptococcus faecalis.* IV. Use of a biphotometer in growth-curve studies. *J. Bacteriol., 84,* 393.

22. Dykstra, M.A., McLaughlin, J.C. and Bartlett, R.C. (1979): Comparison of media and techniques for detection of group A streptococci in throat swab specimens. *J. Clin. Microbiol., 9,* 236.

23. Sanders, E. (1969): Bacterial interference. I. Its occurrence among the respiratory tract flora and characterization of inhibition of group A streptococci by viridans streptococci. *J. Infect. Dis., 120,* 698.

24. Cherry, W.B. and Moody, M.D. (1965): Fluorescent-antibody techniques in diagnostic bacteriology. *Bacteriol. Rev., 29,* 222.

25. Fenton, L.J. and Harper, M.H. (1978): Direct use of counterimmunoelectrophoresis in detection of group B streptococci in specimens containing mixed flora. *J. Clin. Microbiol., 8,* 500.

26. Maitland, H.B. and Martyn, G. (1948): A selective medium for isolating staphylococcus based on the differential inhibiting effects of increased concentrations of sodium chloride. *J. Pathol. Bacteriol., 60,* 553.

27. Freeburg, P.W. and Buckingham, J.M. (1976): Evaluation of the Bacti-Lab streptococci culture systems for selective recovery and identification of group A streptococci. *J. Clin. Microbiol., 3,* 443.

28. Lowbury, E.J.L., Kidson, A. and Lilly, H.A. (1964): A new selective blood agar medium for *Streptococcus pyogenes* and other haemolytic streptococci. *J. Clin. Pathol.*, *17*, 23.
29. Bill, N.J. and Washington, J.A. (1975): Bacterial interference by *Streptococcus salivarius*. *Am. J. Clin. Pathol.*, *64*, 116.
30. Dunkelberg, W.E., Jr. and McVeigh, I. (1969): Growth requirements of *Haemophilus vaginalis*. *Antonie van Leeuwenhoek*, *35*, 129.
31. Vincent, W.F. and Lisiewski, K.J. (1969): Improved growth medium for group A streptococci. *Appl. Microbiol.*, *18*, 954.
32. Patton, A.R. and Hill, E.G. (1948): Inactivation of nutrients by heating with glucose. *Science*, *107*, 68.

Discussion

R.R. Facklam (Atlanta, U.S.A.): Were the broth supernatants checked for antigen?

M. Slifkin: No, though it would have been interesting if we had done so.

I. D. Amirak (London, England): There were 75 % positives from the modified broth; what happened to the other 25 %? Were they cultured overnight and were further attempts then made to group them?

M. Slifkin: In that particular experiment, as in the actual clinical trials, four swabs were taken from each patient. All were culture-positive. One swab was placed on a blood agar culture plate and the other swabs were employed with the experimental broth.

I. D. Amirak: These streptococci were grouped in a time of five hours. Were those that did not group incubated further and tried again, or were they disregarded?

M. Slifkin: No, I just employed five hours as my goal for this group of experiments.

The clinical need and technical methods for the identification of the group B streptococcus

T. M. S. Reid and D. J. Lloyd
Regional Laboratory, City Hospital and Department of Neonatal Paediatrics, Aberdeen Maternity Hospital, Aberdeen, Scotland

The group B streptococcus (GBS) is now established as a major cause of serious neonatal infection [1]. Attempts have been made to identify those newborn infants at risk of developing early onset disease by assessing known maternal and neonatal high risk factors [2]. Current debate centres on the relative merits of antibiotic prophylaxis and immunoprophylaxis. While the exact mechanism of protective immunity in GBS disease remains to be clarified there is already evidence to suggest that early administration of antibiotic is effective in preventing the high mortality from this infection [3–5]. However, in order to rationalize the early administration of antibiotics, it is desirable that laboratories develop techniques for the rapid identification of GBS.

EARLY ONSET GROUP B STREPTOCOCCAL DISEASE

During the period 1973–1978 there were 24,808 live births (≥ 500 g) in Aberdeen Maternity Hospital. There were 27 infants with early onset GBS disease, of whom 10 died, an attack rate of 1.1 per 1,000 live births and a mortality rate of 0.4 per 1,000 live births. Review of the case histories of the 27 infants with GBS disease shows that of the 10 infants who died, 9 were of low birth weight (Table I).

Table I. Mortality due to early onset GBS disease.

	No. of infants	No. of deaths due to GBS	No. of infants < 2,500 g	No. of GBS deaths in infants < 2,500 g
Septicaemia	17	8	10	7
Pneumonia	10	2	2	2
Total	27	10	12	9

ANTIBIOTIC PROPHYLAXIS

None of the eight infants who received antibiotics either prior to or within two hours of delivery died, while there were 10 deaths in the 19 untreated infants (Table II). This suggests that the mortality from this infection can be prevented by early administration of antibiotic.

Table II. Mortality in antibiotic-treated* and untreated infants with GBS disease.

	Total	No. of infants who died due to GBS infection
Treated	8	—
Untreated	19	10

* Mother given parenteral antibiotic during labour or neonate commenced antibiotic treatment within two hours of delivery

Our present policy for the prevention of GBS infection involves the administration of antibiotic not only to all low birth weight infants but also to infants with signs of respiratory distress. However, it is known that the organism is recovered from only 4% of low birth weight infants within two hours of delivery [4], indicating that a significant number of otherwise well low birth weight infants receive penicillin. With this in mind we would suggest that methods should be available for the rapid identification of GBS immediately following delivery. Speed and accuracy are essential since, to be effective in preventing fulminating and often fatal GBS infection, antibiotic therapy must be instituted within two hours of birth.

RAPID IDENTIFICATION TECHNIQUES

Pharyngeal aspirates taken immediately post delivery can be examined for the presence of GBS by co-agglutination or by counterimmunoelectrophoresis (CIE) in addition to standard microscopy and culture [6]. Using these rapid, highly sensitive techniques a positive identification can be communicated to the neonatologist within one hour of delivery. If negative the sample can be tested repeatedly following culture in Todd Hewitt broth. Positives are reconfirmed after culture on standard media for 18 hours.

CIE can be used to detect group B antigen in the blood, cerebrospinal fluid (CSF), pleural fluid and urine of infants [7, 8]. This may be particularly important when there is a need to establish the diagnosis in a sick neonate whose mother may have received antibiotic during labour, rendering subsequent cultures negative.

GBS can also be identified in swabs incubated in Todd Hewitt broth by the fluorescent antibody technique [9]. However, it is important to note that the quality and specifity of commercially available precipitation grouping sera are not necessarily satisfactory for use in CIE or immunofluorescence tests.

ISOLATION, SELECTIVE MEDIA, CARRIAGE RATES

The isolation of GBS from normally sterile body fluids presents little problem but the organism is frequently found in the mixed flora of skin, vagina, urethra, gut or upper respiratory tract. In such circumstances and in epidemiological studies the use of selective media has greatly facilitated the isolation of GBS [10, 11]. The highest isolation rates are obtained by immersing swabs in the selective broth and plating out after 18–24 hours of incubation.

The great variation in GBS carriage rates in pregnancy [12] is to a large extent explained by variation in culture media and methods. No matter how high the carriage rate detectable by selective techniques and repeated swabbing of multiple sites the neonatal attack rates are remarkably uniform [12]. Quantitation of GBS in the vagina [13] may reveal a critical level of colonization possibly detectable by relatively nonselective techniques which can be correlated with an increased risk of acquiring neonatal infection. Equally, there is no doubt that, in the investigation of nosocomial outbreaks which are increasingly being recognized and authenticated by phage typing [14], the use of selective media together with repeated swabbing of multiple sites is essential, if all those colonized are to be identified.

PRESUMPTIVE IDENTIFICATION

While definitive identification of putative group B isolates can only be achieved by serological methods, a presumptive identification may be inferred from the following properties which are characteristic of over 95% of group B strains:

1. *Colonial appearance.* A large (> 2 mm) grey mucoid colony with a small hazy zone of β-haemolysis. Non-haemolytic strains exist (1–6% of human isolates) and, in the absence of selective media, can clearly pose a problem in identification in the routine laboratory, particularly if mixed with normal upper respiratory or vaginal flora.

2. *Pigment production.* GBS will produce a bright orange-red pigment if grown on starch/serum agar. The modified medium described by Islam [15] is particularly effective on primary anaerobic culture, even with mixed growths.

3. *Camp reaction*

4. *Hippurate hydrolysis*

5. *Resistance to bacitracin*

DEFINITIVE GROUPING

The Lancefield precipitin test remains the ultimate reference method for streptococcal grouping but its routine use is being superseded by the introduction of rapid sensitive techniques: slide agglutination [16], co-agglutination [17], counter-immunoelectrophoresis [18] and immunofluorescence [19]. Since there is no requirement for specialized equipment, and the time-consuming antigen extraction step is avoided, co-agglutination is being adopted by many laboratories. As we have indicated, it can be commended particularly to those who provide a service for a busy neonatal intensive care unit.

REFERENCES

1. Parker, M.T. (1979): Infections with group B streptococci. *J. Antimicrob. Chemother.*, 5 *(Supplement A)*, 27.
2. Lloyd, D.J. and Reid, T.M.S. (1976): Group B streptococcal infection in the newborn. Criteria for early detection and treatment. *Acta Paediatr. Scand.*, 65, 585.
3. Steigman, A.J., Bottone, E.J. and Hanna, B.A. (1978): Intramuscular penicillin administratation at birth: prevention of early onset group B streptococcal disease. *Pediatrics*, 62, 842.
4. Lloyd, D.J., Belgaumkar, T.K., Scott, K.E., Wort, A.J., Aterman, K. and Krause, V.K. (1979): Prevention of group B beta-haemolytic streptococcal septicaemia in low birth weight neonates by penicillin administered within two hours of birth. *Lancet*, I, 713.
5. Reid, T.M.S. and Hall, M.A. (1979): Perinatal group B streptococcal infection: a 6 year study. In: *Patholgenic Streptococci*, p. 177. Editor: M.T. Parker. Reedbooks, Chertsey.
6. Slack, M.P.E. and Mayon-White, R.T. (1978): Group B streptococci in pharyngeal aspirates at birth and the early detection of neonatal sepsis. *Arch. Dis. Child.*, 53, 540.
7. Shakleford, P.G. and Stechenberg, B.W. (1977): Countercurrent immunoelectrophoresis (CIE) in group B streptococcal disease. *Pediat. Res.*, 11, 505.
8. Siegel, J. D. and McCracken, G.H. (1978): Detection of group B streptococcal antigens in body fluids of neonates. *J. Pediatr.*, 93, 491.
9. Romero, R. and Wilkinson, H.W. (1974): Identification of group B streptococci by immunofluorescence staining. *Appl. Microbiol.*, 28, 199.
10. Baker, C.J., Clark, D.J. and Barrett, F.F. (1973): Selective broth medium for isolation of group B streptococci. *Appl. Microbiol.*, 26, 884.
11. Gray, B.M., Pass, M.A. and Dillon, H.C., Jr. (1979): Laboratory and field evaluation of selective media for isolation of group B streptococci. *J. Clin. Microbiol.*, 9, 466.
12. Parker, M.T. (1977): Neonatal streptococcal infections. *Postgrad. Med. J.*, 53, 598.
13. Schauf, V. and Hlaing, V. (1976): Group B streptococcal colonization in pregnancy. *Obstet. Gynecol.*, 47, 719.
14. Stringer, J. and Maxted, W.R. (1979): Epidemiological evaluation of a phage typing system for group B streptococci. In: *Pathogenic Streptococci*, p. 262. Editor: M.T. Parker. Reedbooks, Chertsey.
15. Islam, A.K.M.S. (1977): Rapid recognition of group B streptococci. *Lancet*, I, 256.
16. Efstratiou, A. and Maxted, W.R. (1979): Agglutination and co-agglutination methods of grouping streptococci. In: *Pathogenic Streptococci*, p. 254. Editor: M.T. Parker. Reedbooks, Chertsey.
17. Christensen, P., Kahlmeter, G., Jonsson, S. and Kronvall, G. (1973): New method for the serological grouping of streptococci with specific antibodies adsorbed to protein A-containing staphylococci. *Infect. Immun.*, 7, 881.
18. Hill, H.R., Riter, M.E., Menge, S.K., Johnson, D.R. and Matsen, J.M. (1975): Rapid identification of group B streptococci by counterimmunoelectrophoresis. *J. Clin. Microbiol.*, 1, 188.
19. Mason, E.O., Wong, P. and Barrett, F.F. (1976): Evaluation of four methods for detection of group B streptococcal colonization. *J. Clin. Microbiol.*, 4, 429.

Discussion

E. Randall (Evanston, U.S.A.): Is the rapid CAMP reaction, where the CAMP factor is extracted from a staphylococcus and kept as a solution ready for use, beginning to be used in Great Britain or in other countries? We are beginning to use it even in some of the screening cultures done on our neonates who are brought in when there is a non-haemolytic colony that looks suspicious. We drop a portion of the CAMP onto the plate and have the answer within about 10 minutes.

W. R. Maxted (London, England): Yew, we have used it and it can be done in many ways. We have CAMP factor, or rather staphylococcal β-lysin, on strips of filter paper that we can lay on primary or secondary culture plates, or you can streak the staphylococcus down the seeded plate. It is a matter of having a supply of β-lysin.

H. Arvilommi (Jyväskylä, Finland): As a microbiologist, not a clinician, I should like to ask how often these neonatal infections are so difficult to diagnose clinically that microbiological diagnosis helps? How often is the diagnosis of group B streptococcal infection a surprise?

T. M. S. Reid: At present group B streptococci and *E. coli* are the two most common causes of early-onset septicaemia in the newborn. They have a predilection for the small preterm infant who is also at risk of developing the idiopathic respiratory distress syndrome (IRDS). The initial clinical presentation and X-ray findings are similar, making differentiation difficult. While apnoea occurring within the first 24 hours of life is suggestive of infection, infection and IRDS may coexist. It must be stressed that antibiotic administration that is delayed until clinical signs of infection (apnoea, shock and increasing respiratory failure) are evident, may be too late in view of the rapid deterioration seen in severe GBS infection. A rapid bacteriological diagnosis and early administration of antibiotic are essential to reduce the morbidity and mortality from GBS disease.

Quality control data on the identification and serology of streptococci obtained from 500 French laboratories

M. Veron
Central Bacteriological Laboratory, Necker Hospital for Sick Children, Paris, France

This communication is not concerned with our own results from the serological grouping of streptococci but deals with results of experiments relating to inter-laboratory quality control.

These quality control studies have been performed in France since 1975, sponsored by the French Society for Microbiology, with a progressively increasing number of participants, all volunteers, from private or state clinical laboratories. This participation in quality control has now become compulsory for the 3,800 French clinical laboratories working in the bacteriological area.

This report will obviously deal with only some of the problems relating to streptococci.

EVALUATION OF HAEMOLYSIS BY A STRAIN OF *STREPTOCOCCUS PYOGENES*

Many publications have been devoted to studies of the best media and conditions for incubation, in order to demonstrate correctly haemolysis by streptococci. It might be supposed, therefore, that, among the 737 laboratories which responded, many would not have observed the β-haemolytic reaction of the *Streptococcus pyogenes* strain sent to them.

However, the results reported in Table I show that about 88 % of participants gave a correct answer and that there was no significant difference between percentages in relation to the type of blood agar used, or to the manufacturer. The range of percentages of correct results is 95.7 to 78.8.

These results must obviously be interpreted carefully from the technical point of view, because the observed results of a given test may have been modified according to the final diagnosis assumed. This remark is valid for all following sections also.

IDENTIFICATION OF A STRAIN OF *STREPTOCOCCUS FAECALIS*

Table II shows the diagnosis reached by the 533 participants for a strain of *Streptococcus faecalis*, that is to say non-proteolytic and non-haemolytic. In specify-

Table I. Frequency of β-haemolysis observed by 366 routine laboratories for strain 78.F of S. pyogenes.

Blood agar used	Source of medium						Total	
	BioMérieux		Institut Pasteur Production		Home made			
	N^a	$\%^b$	N^a	$\%^b$	N^a	$\%^b$	N^a	$\%^b$
Trypticase-soy agar +5 % horse blood	92	94.6	100	88.0	76	84.2	268	89.2
Trypticase-soy agar +5 % sheep blood	70	88.6	16	81.3	17	82.4	103	86.4
Columbia agar +5 % sheep blood	81	90.1	23	95.7	38	89.5	242	90.5
Miscellaneous media	23	87.0	66	78.8	35	87.5	124	82.3
Total	366	90.7	205	85.4	166	85.5	737	88.1

[a] N = number of results
[b] Percentage of results in which β-haemolysis was observed

Table II. Identification of S. faecalis, biotype faecalis, strain 76.D, by 553 clinical laboratories.

Diagnosis	N	%	Wrong diagnoses
S. faecalis, biotype faecalis, group D	35	6.6	
S. faecalis, biotype faecalis	28	5.3	
S. faecalis, wrong biotype	58	10.9	zymogenes (11), liquefaciens (39), haemolyticus (8)
S. faecalis, group D	184	34.4	
S. faecalis	143	26.8	
Subtotal for correct diagnosis	448	84.0	
Streptococcus sp.	49	9.2	
Streptococcus, wrong species	8	1.5	pyogenes (1), pneumoniae (5), bovis (1), salivarius (1)
Wrong genus	20	3.7	Staphylococcus albus (6), S. aureus (5), Listeria monocytogenes (2), Corynebacterium enzymicum (2), Haemophilus sp. (2), Neisseria sp. (2), Aeromonas sp. (1)
No diagnosis	8	1.5	
Total	533	100.0	

N = number of diagnoses of type shown

ing the diagnosis, participants were obliged to indicate the name of the genus, species and, if possible, the name of the variety and/or serotype.

The diagnoses submitted are indicated in the first column of Table II. Eighty-four per cent of answers were correct, that is to say indicating the exact binomial combination for the genus and species. Among these correct answers there were, however, 10.9% of mistakes in diagnosis of the variety.

Among the answers judged wrong, 9.2% involve a diagnosis of Streptococcus species. This error is probably due more to negligence as regards rules for nomenclature than to a true mistake in diagnosis. Only eight cases in which the species was wrong were recorded and, in about 5% of cases, the participants either gave no diagnosis or a wrong diagnosis. It is in this last situation that the greatest confusion has been observed and, in some cases, this could result from problems during isolation rather than during identification.

Even if the performance of the participants was fairly good, it must be noted that this diagnosis was not difficult.

In addition to diagnosis, the participants had to indicate what characteristics, chosen from a previously established list of 40 characteristics, were used for identification and with what results. The 522 relevant answers have been divided into two groups: group A for the 447 answers in which diagnosis was correct and group B for the other 75.

The frequency of choice of each characteristic in group A has been compared with that in group B. Of the 40 characteristics the percentual usage was statistically

Table III. Frequency of response for some characteristics in the identification of S. faecalis strain 76.D by 522 laboratories (group A = 447, group B = 75).

Characteristic	Correct result	Frequency of use			Frequency of correct results for the characteristic		
		A	B	C	A	B	C
3. Gram-stain	+	99.3	98.7	?	99.8	95.6	?
4. Capsule	−	32.7	49.3	< 0.01	94.5	70.3	< 0.001
6. Small chain	+	76.1	77.3	NS	91.5	84.5	NS
7. Grouping in pairs	+	60.6	60.0	NS	93.0	82.2	< 0.05
11. Blood requirement	−	67.6	76.0	NS	64.9	45.6	< 0.01
14. Growth at 45°C	+	16.1	8.0	NS	91.7	33.3	?
15. Thermotolerance (30 minutes, 45°C)	+	36.9	17.3	< 0.001	96.4	84.6	?
16. Growth on Chapman	−	34.5	48.0	< 0.05	66.2	69.4	NS
17. Growth on NaCl 6%	+	34.9	20.0	< 0.05	94.9	46.7	?
19. Bacitracin sensitivity	−	49.2	57.3	NS	94.1	79.1	< 0.01
20. Optoquine sensitivity	−	52.6	56.0	NS	100.0	83.3	?
24. Production of catalase	−	72.0	56.0	< 0.01	94.7	95.2	?
32. Esculin catabolism	+	89.9	52.0	< 0.001	98.8	84.6	?

A = Frequency (%) for laboratories with a correct diagnosis (total = 447)
B = Frequency (%) for laboratories with a wrong diagnosis (total = 75)
C = Probability, using the Chi-square test of significant difference: NS (= not significant) = $p > 0.05$; ? = sample size too small

significantly different for eight, of which six are shown in the simplified Table III. The two other characteristics are the presence of spores and the fermentation of L-arabinose. These two last characteristics, although without usefulness, were chosen more frequently in group B. In contrast, some tests such as thermotolerance, catalase production and esculin catabolism, which are essential for the diagnosis of *S. faecalis*, were chosen much less often in group B. The significance of these observations, considered as a test of knowledge, is worth thinking about.

Table III also shows the frequency of correct results for some characteristics chosen for identification in group A and group B, respectively. The participants who arrived at a wrong diagnosis more often found wrong results for at least four characteristics out of the 13 cited in Table III. It may readily be supposed that the mistakes in these tests explained the incorrect diagnosis.

SEROGROUPING OF A STRAIN OF *S. PYOGENES*

Another problem tested by the participants in the quality control system was the

Table IV. Serological grouping of strain 78.F of S. pyogenes.

Lancefield group	Number of diagnoses
A	811 (86.8 %)
B	19 (2.0 %)
C	14 (1.5 %)
D	9 (1.0 %)
F	1 (0.1 %)
G	3 (0.3 %)
H	5 (0.5 %)
Other	0 (0.0 %)
Unspecified	72 (7.8 %)
Total	934 (100.0 %)

Table V. Serological grouping and biogrouping of S. pyogenes strain 78.F.

Technique*	Group A or S. pyogenes (%)	Other groups (%)	Total number
Immunoprecipitation	86	14	100
Immunofluorescence	94	6	18
Co-agglutination (Phadebact® Streptococcus Test)	87	13	228
Counterimmunoelectrophoresis	90	10	10
Api-Strep® system	85	15	307
Bacitracin test	87	13	250
Unspecified	95	5	21
Total	87	13	934

* No statistical difference

grouping of a strain of *S. pyogenes*. A summary of the results is shown in Table IV. It emerges that 86.8% of participants reached the correct answer, that is to say Lancefield group A. Only 51 answers related to different groups. On the whole, therefore, these results may be considered satisfactory. Table V shows that the techniques used for grouping had little effect on the correctness of the answers, since the frequencies of correct results are all greater than 85% and do not differ significantly.

ANTISTREPTOLYSIN O TITRATION

The last question considered here concerns the titration of antistreptolysin O. The same sample of serum was sent to each participant, without insistence on any particular method for titration.

Histograms of the results obtained by the 494 participants appear in the Figure, on the left in relation to all results and on the right in relation to results observed with the three materials most often used. Review of these histograms allows the conclusion that the results from the different materials used do not vary much and that the average lies between 200 and 400 units. Computation confirms this impression, as the calculated averages and standard deviations of the titres for each material respectively are (logarithm to the base 2) 1.69 (323 units) $\pm$ 0.79 for 'Bio-Mérieux', 1.82 (353 units) $\pm$ 0.71 for 'Institut Pasteur Production' and 1.57 (295 units) $\pm$ 0.61 for 'La Technique Biologique'. The results obtained with the three methods do not differ significantly. The confidence intervals calculated are consistent with an acceptable precision for the titration, whatever the material used. However, having regard to this range of variation, it may be stated that it is pointless to adopt serial dilutions closer than 1 : 2.

To conclude, I hope to have shown, with results concerning streptococci only, that the interlaboratory quality control system is an important tool for discovering how the techniques well established in reference laboratories may be applied in any clinical laboratory.

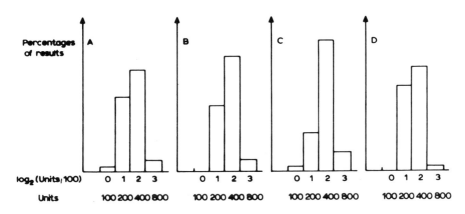

Figure. Histograms of frequencies of results obtained by 494 laboratories for antistreptolysin O titration: A: Overall results (494 responses); B: Results obtained with 'BioMérieux' material (147 responses); C: Results obtained with 'Institut Pasteur Production' material (40 responses); D: Results obtained with 'La Technique Biologique' material (286 responses).

Short presentation

A.-S. Malmborg
Clinical Microbiological Laboratory, Huddinge University Hospital, Huddinge, Sweden

I should like to report a slight technical improvement in the Phadebact® Strepto-coccus Test. I am very glad to have the opportunity to do this at this meeting in the presence of Dr Maxted.

We have used the Phadebact® Streptococcus Test co-agglutination technique for a long time and we try to give the results to the clinicians as quickly as possible, There-fore, we try to perform the agglutination from the primary plate. If five or six pure colonies are seen, they are pretreated for two hours and then agglutinated. For pre-treatment we have, up to now, used trypsin but we are now trying to use Maxted enzyme which is a mixture of streptomyces enzymes. Dr Maxted published in this connection several years ago.

As a reference method we have used incubation in Todd Hewitt broth for 24 hours.

Results of a comparison between pretreatment with Maxted enzyme or trypsin for two hours and the reference method, Todd Hewitt broth, for 24 hours are shown in Table I. It is a very small trial, with strains from only 20 patients in each group. Pretreatment with Maxted enzyme appears to be better than trypsin.

Agglutination was stronger and easier to read with pretreatment using Maxted enzyme (Table II). The strong reactions, + + + and + +, are more frequent with Maxted enzyme than with trypsin.

There were also fewer reactions if Maxted enzyme was used for Group D and the false reactions that were obtained were much weaker than with trypsin pretreat-ment.

In conclusion, we have found that it is better to use Maxted enzyme for pretreat-ment and the technicians in our laboratory are pleased that it is now used routinely.

Table I. Comparison between three Phadebact® Streptococcus Test procedures for serological grouping of β-haemolytic streptococci: positive reactions.

	Number of strains tested	Number giving positive reactions		
		Maxted enzyme (2 hours)	Trypsin (2 hours)	Todd Hewitt (24 hours)
Group A	20	20	18	20
Group B	20	18	15	20
Group C	20	19	19	19
Group G	20	20	16	16

Table II. Comparison between three Phadebact® Streptococcus Test procedures for serological grouping of β-haemolytic streptococci: degree of co-agglutination.

| | Number of strains tested | Degree of co-agglutination | | | | | | | | | | | | | | |
| | | Maxted enzyme (2 hours) | | | | | Trypsin (2 hours) | | | | | Todd Hewitt (24 hours) | | | | |
		+++	++	+	−	Other groups	+++	++	+	−	Other groups	+++	++	+	−	Other groups
Group A	20	18	2			1	10	7	1	1		18	2			
Group B	20	3	14	1		2	1	7	7		5	20				
Group C	20	9	8	2		1	2	11	6	1		16	3			1
Group G	20	8	11	1			2	9	5		4	14	2			4

Summary

R. R. Facklam

Some of the presentations today reminded me of a term used to describe the streptococci: they are one of the more ubiquitous bacteria. This fact presents clinical microbiologists with not only a challenge regarding accurate identification of the streptococci, but also a challenge in relation to judgement of the importance of finding the various streptococcal species in clinical samples. The relationship of the presence of streptococci in clinical specimens to the disease state is not always known. Clinical microbiologists need to know taxonomy and to have a basic knowledge of medicine in order to help understand the relationship of the various streptococcal species to different diseases. Many of the laboratory techniques, such as serological grouping procedures, offer the clinical microbiologist a taxonomic tool and a method of identifying pathogenic species.

There are many levels of competence in laboratories that deal with the streptococci and the infections they cause. There are large reference laboratories that perform sophisticated serological and physiological tests that identify not only the streptococcal species, but the serovars and biovars of species. However, because of their complexity and cost the techniques used in these laboratories, although conventional, are not applicable in all laboratories. Probably the most widely used techniques for identifying the streptococci are the presumptive identification techniques. There is no doubt that presumptive test schemes are accurate and provide sufficient information in relation to properly managed patients with streptococcal infections. These procedures, however, do not yield sufficient information regarding the epidemiological or the pathological nature of the streptococci. The development of simple, rapid tests that will provide us with more specific identification will, therefore, be very useful.

No less than eight of our colleagues have presented data today on four recently-developed techniques that will improve the laboratory identification of streptococci. The majority of reports regarding the co-agglutination test were very favourable. A word of warning was sounded, indicating that specific conditions must exist before the test is reliable. The opinion of most investigators using the co-agglutination grouping reagents is that they are more sensitive and much more simple to use than conventional grouping reagents, when used under defined conditions.

The majority of reports concerning the use of the latex agglutination grouping procedures were also very favourable. Although these reagents have not been available as long as the co-agglutination reagents, most investigators reported comparable results. The latex agglutination reagents are probably not quite as sensitive as the co-agglutination reagents but, like the co-agglutination tests, they provide the clinical microbiologist with a more accurate procedure for identifying the streptococci than presumptive test schemes.

The nitrous acid extraction technique would seem to offer the clinical microbiologist a very useful tool to identify the streptococci serologically. However, there is insufficent information at present to allow its usefulness to be judged.

The most recently developed system and the one on which we have the least amount of information is the API® biochemical identification scheme. We heard one favourable and one unfavourable report on this system. The conclusion is that not enough information is available to make a final judgement about the usefulness of the API® system. Perhaps future experiments will resolve this problem.

I would like to conclude this meeting with a note of optimism. I feel that several of the techniques described recently provide not only clinical microbiologists but microbiologists in other disciplines with useful tools to improve the identification of streptococci. Serological grouping of the β-haemolytic streptococci is the most definitive method we have for such identification. The development of the rapid agglutination tests makes serological grouping of the β-haemolytic streptococci available to laboratories at all levels of competence. The more definitive the identification, the more we will understand about the streptococci and the role each species plays in infection.